W9-BPQ-178

Reimagining our energy system offers the potential of a win-win-win future: low carbon energy, more resilient communities, vibrant local economies. What's more, it's possible and it's already happening. In these pages, Howard Johns paints a spellbinding picture of the revolution happening around us, and the part you could play in it.

Rob Hopkins
founder of the Transition movement

This is an important book for those wanting climate and energy solutions rather than just criticisms. Its survey of decentralised energy around the world is extremely valuable.

Stephen Tindale
co-founder of Climate Answers

We must invest in renewable energy now to avert the worst effects of climate change. Howard Johns' *Energy Revolution* makes a compelling case for repowering our communities, putting an end to dirty power and the monopoly of the Big Six, and easing fuel poverty. Essential reading for anyone who wants to understand the energy industry and how we can transform it.

Caroline Lucas
Member of Parliament (MP) for Brighton Pavilion
since 2010 and Britain's first Green MP

Howard Johns ably explains the imperative of switching to renewable energy and presents stories from around the world that demonstrate how this emerging industry can create attractive investments as well. An inspiring and grounded book.

Ben Goldsmith
CEO, Menhaden Capital

ENERGY REVOLUTION

Your Guide to
Repowering the Energy System

HOWARD JOHNS

Permanent Publications

Published by
Permanent Publications
Hyden House Ltd
The Sustainability Centre
East Meon
Hampshire GU32 1HR
United Kingdom
Tel: +44 (0)1730 823 311
Fax: +44 (0)1730 823 322
Email: enquiries@permaculture.co.uk
Web: www.permanentpublications.co.uk

Distributed in the USA by
Chelsea Green Publishing Company, PO Box 428, White River Junction, VT 05001
www.chelseagreen.com

© 2015 Howard Johns
The right of Howard Johns to be identified as the author of this work has been asserted
by him in accordance with the Copyrights, Designs and Patents Act 1988

Typeset by Emma Postill

Cover design by Harriet Lamb

Printed in the UK by CPI Antony Rowe, Chippenham, Wiltshire

All paper from FSC certified mixed sources

The Forest Stewardship Council (FSC) is a non-profit international
organisation established to promote the responsible management
of the world's forests. Products carrying the FSC label are
independently certified to assure consumers that they come from
forests that are managed to meet the social, economic and ecological
needs of present and future generations.

British Library Cataloguing-in-Publication Data
A catalogue record for this book is available from the British Library

ISBN 978 1 85623 197 8

Contents

PART III   Your Guide to Making It Happen

PART IV   Step Forward in Hope

Foreword

Ninety-seven percent of the world's scientific community think that climate change is caused by human activity whilst one in four Americans believe that its effects are exaggerated. Clearly, we have a cognitive divide between the evidence and public perception. An entire reframing of social attitudes towards the impending ecological crisis and the dramatic shift away from burning fossil fuels, a deeply entrenched habit that we have to kick, is essential.

Fortunately, it is totally possible to move all our energy systems over to renewables but the crux is that in doing so we need an energy revolution, as Howard Johns explains, and we need it fast. This revolution will not only involve technology, it will require that we, in our villages, towns and cities, become a part of that process and get organised. The key to efficient renewables is the creation of de-centralised systems in which local people are involved. The opportunities for both technological and social innovation are huge.

This book is full of case studies from all over the world that demonstrate how communities have set up their own generation systems and energy companies, and make dramatic shifts in short periods of time. In the developed world where we are passively trained to expect (dirty) power 24/7 at the flick of a switch, an energy revolution will create jobs, reduce greenhouse gases and tackle fuel poverty, and is an opportunity for more active community engagement. In the developing world, renewable technologies are already allowing young people to gain a better education, reduce deaths and illness from polluting kerosene lamps, and enable people to create income by working or studying in the evenings.

Howard uses tested examples from all parts of the globe and shares his insights and experience from two decades of work both project managing installations and setting up community generation systems. He makes a powerful case for energy revolution, proving that effective renewable energy systems are not only appropriate in every country, they are the future, the only future we can have. We can all be energy entrepreneurs.

Dr Jeremy Leggett
Green energy entrepreneur, author and activist;
founding director of Solarcentury, the UK's largest solar installation company;
founder and Chairman of SolarAid and Chairman of CarbonTracker

For all our relations

Author's Note

It was with a backdrop of massive challenges in my work life that I decided to write this book. The UK government made sweeping changes to the mechanisms supporting the solar industry of which I have been an active part for the last 15 years. The company I had spent 10 years lovingly building was now in peril and I began the process of closing offices and sacking friends in order for it to survive. A depressing and infuriating task and I really needed something positive to focus on. Finding and telling great stories about people's successes in pioneering community renewable energy around the world was my light relief and source of hope. This is part of what you hold in your hands.

I first studied climate change in my energy technology degree in the early 1990s, and quickly became an angry young man because of it. Consequently, I lived for quite some time on environmental protest sites around the UK, and was even evicted from my treehouse home whilst trying to prevent the digging of an opencast coal mine in south Wales.

Protesting has a certain appeal when you are angry, but my days of saying no to things soon were not enough to satisfy my desire to see meaningful change. So with a shift in direction, I set about building a company that attempted to deliver that change, making it easy for people to say yes to a more positive future.

My first attempt failed, but eventually I achieved success with Southern Solar, and later with Ovesco, which came out of Transition Lewes. For me it has been quite a journey with lots of ups and downs, many challenges and great moments along the way.

There have been many excellent people who have helped me in the writing of this book that I would like to acknowledge.

Firstly, there are lots of brilliant folks and some personal heroes of mine interviewed in this text. I was overwhelmed by the generosity with which people gave their time, insight and experience for this project. As you will read about them yourselves, I won't mention them all here. However a few people I interviewed are not actually quoted in the book: Rt Hon. Greg Barker, Jigar Shah and Sir David King all added valuable insights. Thank you. I recorded the interviews in person or on the phone over the course of a year and have to thank Lucy Hemming for painstakingly transcribing many of them.

Many people have edited, advised, introduced, tidied, proof read, corrected, and reworked sections of this book. For me they are a bit of a stellar cast, many of whom are experts in their field. So a huge heartfelt thanks to Liz Mandeville, Annie Townend, Camilla Bausch, Chris Sowerbutts, Pete Grech, Martyn Williams, Paul Dorfman, Jon Halle, Nick Rouse, Chris Rowland, Charlotte Webster, Anna Guyer, Sebastian Berry, Tom Peterson, Anna Leidreiter, Jan Erik Nielsen, Arran DeMoubray, and Uwe Trekkner.

In turning this into the book you have in your hand, I have to thank Maddy and Tim Harland and the team – Tony, Emma, Rozie and John – at Permanent Publications. Thanks for believing in the book and being patient with me in the process! Thanks also to Harriet Lamb for the cover.

For me, personally, none of this would have been possible without a huge cast over the last 15 years who got on board with the various missions and worked so hard to make my successes possible.

Thanks to the hundreds of great people that have worked with me at Southern Solar. I am grateful for all the good work we have done – and especially for the great fun we have had together in doing it – despite the very hard bits. I think we have done it with style and I am very proud of the many homes, schools, communities and businesses we have solarised so far!

Thanks also to the many people I worked with at the Solar Trade Association where I acted as Chairman for nearly five years. I learnt an awful lot in the process, and I do think we had an impact.

Thanks to the good people of Lewes that came together in Transition Lewes, led by the determined and inspirational Adrienne Campbell. Thanks for bringing so many people together in hope and helping to create meaningful change in our town.

Out of Transition came an energy group and eventually Ovesco. I am really grateful to all who played their part in this but particularly want to acknowledge my fellow Ovesco directors: Dirk Campbell, Liz Mandeville, Chris Rowland, Nick Rouse, Paul Bellack, and Ollie Pendered. It has been one hell of a journey to make Ovesco happen and I am really proud of what we have achieved so far and grateful to you all for giving so much of yourselves in the process. I look forward to what comes next.

Many people have supported me personally over the years and I want to mention a few of them here. Thanks to John Kirk, Nick Carling, Jez Hughes, Rob Whitty and Roger Ross for all the love and support.

I am very lucky to have amazing parents who have supported me all the way, and a father that has not only supported, but got involved too. I feel very privileged to have worked with my Dad on my projects, have learnt a lot and grown a whole load from his input. A pretty rare gift it seems.

And lastly I would like to thank my family, who have put up with years of me being stressed and overworked by all of the above! Most recently they have put up with a year and a quarter of me being extra tired whilst writing this book in my evenings and at night when really I should have been sleeping. I am very lucky to have two awesome sons, Orin and Luca, an amazing wife Pippa and to be sitting here writing this with my three week old daughter Wren in my arms. What a blessing.

To write about energy at this moment of change is somewhat of a challenging task – things are moving so fast and constantly need updating. I felt it was important to create a snapshot of this amazing moment, and the huge opportunity and challenges that we face. Despite all of those challenges, I feel more strongly than ever that we can create a better future for ourselves on this beautiful planet if enough of us come together, get engaged and get on with it. I hope that in writing this book and sharing so many sources of hope, as well as some practical steps, that more people will be inspired to play their part in making the change we need.

Howard Johns
August 2015
Sussex, UK

www.energyrevolution.solutions

Introduction

It is time to build the energy system that we need for the 21st century and to stop sitting and waiting expecting someone else to do it. I have written this book to share my experience of trying to do just that, and of what I have seen in other communities both here in the UK and globally. This new system will be one that harvests the energy we need for our lives directly from around where we live, from the natural resources we have available: sun, wind and water. It is one that can be owned and will be operated locally, that done well will give communities control over this crucial area of life. It is one that has the power to democratise energy, taking it out of the hands of the large corporations and putting back into the hands of the communities it serves.

In developed nations we need to build the new energy system on the foundations of what already exists, which on the face of it works, but is starting to crumble around us. It is a system that concentrates power into the hands of a few corporate entities, whilst the rest of us are at the end of the pipe and our only option is to pay the bill. It is a system that has massive costs that are not accounted for as it extracts and uses fossil fuels with no recompense to the damage to climate and communities along the way. In developing nations this new system will help people to leap frog the old model and to create an energy system matched to local needs and resources.

In many ways it is turning the clock back to a time when all of our energy needs were met directly from the resources around us, when there was a windmill and a watermill in every town. We now address this challenge, however, with all the knowledge, skills and technology we have developed during the industrial and digital revolutions. There are many communities that have already embarked on this journey and they have created maps that show us the first steps to take.

We live in an unprecedented time of challenges, but also of opportunity. The challenges we face today are unlike any previous generations have faced: huge, complex and seemingly unsolvable. We have climate change, resource depletion, fossil fuel dependency and price volatility, and the concentrated corporate powers of large energy companies to name a few. There are many barriers – real, perceived and manufactured – that are preventing us from taking meaningful action to change our societies so that they come into balance with nature as we know we must. We are simply not moving fast enough to address the massive threats posed to humanity and life on earth by climate change and resource depletion. In developed countries, in our comfortable consumer lives, we seem to be sleepwalking into a whole new and seemingly far more unpleasant reality. For many in developing nations, the impacts of anthropomorphic climate change are already starting to have very severe effects on already challenging lives. Life on this beautiful planet and the future of our children is at stake.

I am assuming, however, that readers of this book do not want to spend time debating the facts and disastrous impacts of climate change. It seems rarely a month passes now without another 'once-in-a-century' weather event threatening communities in different areas across the world. The prognosis is not good. The Intergovernmental Panel on Climate Change (IPCC) reports say all you need to know on this subject, and if you are unconvinced I recommend you don't bother with this book but go and read them instead. It is sobering reading with not much hope to be found within.

For those of us in developed nations, our use of fossil fuel energy has driven the creation of our societies and structures and pervades every aspect of our lives, from food production to pharmaceuticals, construction to conflict, entertainment to education. We have created an incredibly complex system of distribution and extraction to keep this life-blood flowing, at seemingly all costs. I am not going to dwell on the arguments of peak oil as there are many others more qualified than me to expand on it and already many good books on the subject. All I will say is that fossil fuels are finite, they seem to be available in less and less places, and what is more they are harder and harder to get out of the ground. The net result of all of this is that they are getting more and more expensive, and this makes it hard to run our businesses and homes in the way we have become accustomed.

The structures we have created to power our communities are like the dragons of our times: breathing fire and spewing smoke. The irony of the situation is that whilst many of us are concerned about their presence in our midst, we all have a guilty secret – that we are the ones feeding these dragons – every month we pay our bills and every time we fill our cars. There seems little we can do to stop that.

This book is not about slaying dragons, but about owning our involvement in their success, finding a way to deprive them of the food we knowingly give them, and thus freeing ourselves from their shadows. It is a call to you to take up the challenge of building the new infrastructure we urgently need, in your town, business, or farm and with your local community.

Maybe you already do your recycling, buy local food, turn down the heating, walk to work and all those other important steps – but you don't feel like it is making much of a difference. It is hard to feel that our small actions have large effects when we are often bombarded with the magnitude of the problem. For me in the UK, I am very aware of the argument "that we are only responsible for 2% of emissions globally – and what difference can we possibly make".

This is a book of hope, as personally I believe we all make a difference and what is more we have all the tools we need to bring our societies back into balance with the planet in my life time. We just need to get on organising ourselves and use them. Counter to what our politicians often seem to suggest, we cannot afford to wait for others who are 'worse offenders' than us to make the first move, or we will never do anything. Instead we must get on and lead, building a future for our children whatever the barriers. Many communities around the world have already walked a fair way on this path, and we have them to call on for inspiration and guidance, but now we need many more around the world to follow suit for the changes to gather speed and for the benefits to start to flow.

These benefits will be huge. Money flows out of our communities every time we switch on the light and turn on the heating or aircon. This money could stay local and help our communities to be more resilient in the face of these changes. Many people in our societies are struggling to find any work yet there is meaningful work aplenty in facing these challenges and solving these problems. A whole new crop of social entrepreneurs are stepping forward to lead the charge and many more are needed to join them. This book is intended to be a companion on that journey.

I don't presume to have all the answers to the many complex problems we face, so those who want a big theory to answer the multitude of problems may be disappointed. But what I offer here is one route that I hope will quickly get communities on the journey of developing, building and owning their own renewable energy power stations. Starting with small first steps, the process I describe in this book will help your community to develop a new relationship with its energy supply. It will be the beginning of a journey on which your community will take increasing responsibility and, importantly, ownership for the energy it needs – with all the challenges and rewards that brings. It is a path that I have had success with myself and that can definitely be replicated by others. We cannot hope to meet all of our needs in our first steps, but that is no reason not to act. My hope in writing this book is to encourage people to take the all important first step, the one that is often the hardest to take. Once you are on the path the road will unfold in front of you – as there are many opportunities once you understand the business model and have some successes under your belt.

My journey in community energy started in Brighton, UK, in the late 1990s and was quite frankly an absolute disaster. I tried to set up a company called 'Community Energy Systems' and naively went out to recruit others to join this noble cause, telling people that I wanted to set up a community-owned power station. I gathered a small team, but we were met with blank looks and lukewarm responses. It soon became apparent that the wider population did not share my enthusiasm for this task – in fact most people seemed to think I was somewhat crazy. Sadly, my little band soon ran out of steam and we went our separate ways, leaving me perplexed as to why this great idea had fallen flat.

It took me a while to find my way and, as a slightly easier alternative, I set up a solar company in 2002, Southern Solar. I spent the next 10 years building this into a company based in seven locations across the UK with a team of about 90 employees. In 2007, I saw an opportunity to rekindle the original community energy idea I had started with, when the Transition Town of Lewes was 'unleashed', and I ended up in a group of people in the town looking at how we would address our energy needs locally. When someone suggested we should build ourselves a wind turbine, I knew that the idea had come of age and the time was ripe to make it happen.

If you have picked up this book, perhaps this is your moment.

If not you, who? If not now, when?

PART 1

From Edison to Fracking

CHAPTER 1

GLOBAL PYROMANIA

Over the last few years I have stood up in front of many community groups to talk about energy and environment. The thing I normally open with is a statement – if you want to know why your electricity is so expensive have a look at the system used to create it. Sixty percent of the energy that is put in at the start goes straight up the chimney as waste heat.

Every time we turn on the light switch, start our cars, turn on the TV, fires are burning to make it all happen. Huge plumes of CO_2-rich smoke follow us in all we do, but in many countries we have just got better at hiding it. Next time you take a ride in your car, turn on your boiler, turn on the light, visualise the fire that burns to make it happen. Next time you are sat on the motorway in traffic, look about you at all those other cars, and imagine the fire neatly hidden under the bonnet, burning hot.

In virtually every culture and mythology there are stories of monsters and dragons that prey on our communities and threaten the lives of the people. In the 21st century it seems like we have found a new set of dragons in our midst and once more we are feeling the heat from their fiery breath. Perhaps this is true of every story through history, but it is definitely the case that at this time these dragons are of our own creation. Our power stations needlessly spewing waste heat and pollution have become like those monsters. In fact not only did we create them, but we are all conspiring to keep them alive, even though we know that they are poisoning the future for our children here on Earth. As far as I know we don't have another planet to move to right now – so we really do need to work this one out.

The energy companies that provide electricity and gas to your home, and work on our behalf to drill, mine and gather the resources to power the energy infrastructure we rely on, I am sure were built with good intentions. The people who work in them probably believe in the service they provide and, as humans like me and you, are most probably aware of the scale of the problem, though, like the rest of us, perhaps don't know how to take any action to change. These large corporations are like the very supertankers that serve us our crude oil – very large, and slow to turn and stop. There are people within them trying to change things but the inertia is immense.

All of them have invested heavily on building the infrastructure we rely on: gas drilling wells, mines, processing facilities, storage, distribution, shipping, power generation plants, networks, transmission. It all connects to our houses, so we are warm and have light and power. Apart from seeing the odd cooling tower and a few pylons the closest we come to this is actually switching on the lights, turning on our heating or putting fuel in the car. But what lies behind this is a vast industrial system sending billions of tonnes of carbon back into the atmosphere every year. Billions of tonnes of carbon that have been locked away in the crust of the earth for hundreds of thousands of years but have been released again in the last couple of hundred.

The global energy system is becoming more extreme in its methods to feed our appetite. It is a system that has become the dominant economic force globally. Businesses that are larger than countries and that have more power and resources than governments are driving forwards with their agenda of extracting their reserves and making a return for shareholders – despite the evidence of the harm their business models are causing. Driving away down a blind alley, and we are all being dragged along in the slipstream. We are all part of their success story, with every gallon of gasoline we pump; every time we flick the switch, we put money in their hands.

It is painful to look at this stuff straight on and not to feel hopeless and powerless. But I encourage you to consider the things that maybe you don't want to consider. Global warming is happening and it is the biggest threat our species have ever faced. Our current trend for extraction of fossil fuels is quite frankly madness. Nuclear energy in its current form is nothing more than a huge toxic and financial legacy for future generations.

Sometimes we just have to steel our nerves and face down the stuff we don't want to face, and the first section of this book is a bit like that. This is a book about hope and solutions, about action and community, about the power of small groups to change everything. First however, you need to immerse yourself in the problems, for a brief period, to give you the knowledge and drive to work on the solutions. As one of my inspirations says to me, "in the darkness you find the gold". I encourage you to dive into the darkness and find the gold within yourself. Take a whistle stop tour from Edison and Tesla, to fracking and tar sands, via energy that will be 'too cheap to meter', islands sinking beneath the oceans, politicians who seem powerless to effect change and straight on to the next big crash – unburnable assets. Step on board for what could turn out to be the worst fracking nightmare of a future, unless we realise where we are and take some action. But don't worry, that is what the rest of this book is about. Deep breath.

CHAPTER 2

FLICK THE SWITCH

Follow the wires back from the switch, to the fuse box and out of your house, down the street and across the fields, all the way to the power station. An amazing system built over a century with last century's technology and thinking at its heart. Fire in the heart of it making steam to turn the turbines that make the electricity that flows into our homes. Big power stations that are often sat near the coal mines that feed them to make it possible to put in the huge amounts of coal required. Huge plants sat miles away from where the energy is needed.

My grandfather was shift chief in a power station in Southampton during the Second World War, and my father describes his boyhood memories of being in the control room, sat on the shoulders of one of the workers as he pulled the massive levers that shut off power to sections of the city. In those days the electricity grid that now spans Europe and the US was still under construction. When there was too much demand for electricity from the residents of Southampton, they would have to turn some of them off to keep the lights on for everyone else. In those times of challenge it was common to have rolling blackouts on the system as demand outstripped supply. In many parts of the world this is still a reality. In others there is no access to the grids, and no access to electricity at all.

Electrical supply has come a long way in the last 150 years, in some parts of the world at least. Innovation around electricity stormed ahead in the 19th century, and in the 1870s renowned inventor Thomas Edison built the first generating plant in the US in New York City.[1] The first public supply of electricity was in Godalming in the UK in the early 1890s, where public lighting was provided by a hydropower scheme. Lighting was one of the key uses of electricity in the early days of its use – it often replaced gas for street lighting. In fact in Godalming there was opposition to this innovation from some quarters including the local mayor who allegedly had shares in the local gas company.[2]

By 1920 there had been an explosion of power generation and electrical supply all over the US and Europe, a huge entrepreneurial energy

---

[1]   Emergence of Electrical Utilities in America, http://americanhistory.si.edu/powering/past/h1main.htm
[2]   First Public Electricity Supply in the World, www.gracesguide.co.uk/First_public_electricity_supply_
        in_the_world

transformation as new companies and municipalities built generation capacity such as steam turbines and hydro turbines, and distribution systems. In the US the electrical output from utility companies grew massively from 5.9 million kWh in 1907 to 75.4 million kWh in 1927, with the costs dropping 55% in the same period.[3] Most of the first power stations would only transmit power for a mile as they were producing 110V direct current. The use of alternating current changed this. The market that was pioneered by Edison was taken over by Tesla's invention of alternating current for the easy transmission of energy across distance. The system that has now been adopted worldwide was dreamed up in the last few years of the 19th century by a man who was born in the 1850s in what is now Croatia, the son of a Serbian Orthodox Priest.[4] Incredible really, a testament to the transformative power of ideas.

The speed with which the systems were rolled out meant that there was a range of different designs of systems and operations. Often more than one generation station and transmission system was located in one city. As the alternating current distribution system became available, consolidation occurred with larger companies buying up smaller generators and simply distributing from larger more cost-effective stations. Invariably the system started to be regulated, by various acts of Parliament in the UK, and Congress in the US. Generally these acts sought to create a standard for the market and in some cases make access to energy easier for a wider group of people. In some cases they enabled the local municipality to take ownership of the system in their area. The UK Electricity Supply Bill of the 1920s set about creating a series of Regional Electricity Boards, which began nationalising and rationalising the generation assets. It also paved the way for the creation of the national distribution system – the 'Gridiron' as they called it then – now called the national grid. It actually took many years to build this system – a 30-year project with huge investment from the government.

At first electricity was a novelty, and people flocked to Pearl Street in New York City to see it. Quickly the benefits of using electricity to run motors, fans and therefore factories was understood and the demand for electrical power spread into the day as well. Many of the generation stations feeding into this system were coal fired, running steam turbines to generate the power.

Fast forward to today, and there are electrical distribution systems and generators spread all over the earth. This mega energy system generated 21,000 billion kWhrs in 2011 globally.[5] Anywhere there is a density of human population you can often find large power stations and distribution systems. In 2012 coal-fired power stations accounted for some 41% of the total electricity generated[6] with some 7,000 power stations worldwide. Nearly 8,000 million

---

3   Emergence of Electrical Utilities in America, http://americanhistory.si.edu/powering/past/h1main.htm
4   Tesla Biography, www.teslasociety.com/biography.htm
5   International Energy Statistics, www.eia.gov/cfapps/ipdbproject/IEDIndex3.cfm?tid=2&pid=2&aid=12
6   Coal and Electricity, www.worldcoal.org/coal/uses-of-coal/coal-electricity

tonnes of coal was produced around the world in 2012, with about 75% being used for steam production. Nearly half of all the coal produced globally is used in China and according to the International Energy Agency (IEA), 25% of all human carbon emissions come from coal-fired power stations. Other forms of electricity generation come from gas-fired power stations, nuclear, oil fired, and of course renewables with the mix varying greatly from country to country, depending on the resources available locally, and the politics.

So what happens when you stick a tonne of coal into a power station? You turn it into a powder to get the best combustion (if you are running the most efficient coal plant). It burns in a furnace to heat water to turn it into steam. The steam then turns a turbine, which is used to produce the electricity. Essentially this is no different from what was being done 150 years ago. The crazy thing is we are still running this system with very low levels of efficiency. For every tonne of coal put in the furnace, around 60% of the raw energy is lost straight up the chimney as waste heat – and in many cases more than that. The efficiencies for gas-fired power stations are better, but the average efficiency is still really low. I ran the numbers for the UK, based on the Department of Energy and Climate Change's statistics, and the average efficiency of the fossil fuelled part of the electricity system was around 36%.

In the UK our biggest coal power station is Drax, originally built to take advantage of the Selby Coalfield. This single site produces around 7% of UK electricity supply and is marketed as one of the 'greenest' power stations in the UK. It is becoming our 'greenest' power station as they are starting to burn biomass in it and not just coal – so essentially replacing the fossil fuels with ones that can be re-grown – trees. In 2013, 8.5 million tonnes of coal was poured into this monster, as well as 1.6 million tonnes of biomass. I have no idea what 10 million tonnes of fuel would look like, but it sounds like a huge pile. Much of the coal it uses comes from abroad – I have been struggling to find out exactly where, but I believe the bulk of it comes from Poland. It is all delivered in trains, each car carrying 75 tonnes of coal. In 2013 23,000 tonnes of coal was burned every day, equating to about 310 coal wagons; that's a pretty big train. All of this input, and what comes out the end? In 2013 Drax produced 23,000,000 tonnes of CO_2 (with just under 3,000,000 of these being from biomass so not normally counted), 31,700 tonnes of sulphur dioxide, and 39,200 tonnes of nitrous oxides.[7]

On the global scale Drax is actually not the biggest power station or emitter, despite being the UK's largest single source of CO_2 emissions. Topping the list of big emitters (in 2009) is the Taichung power plant in Taiwan, weighing in at 36 million tonnes of CO_2. Selection of the other big hitters is as follows: the Poryong Plant in South Korea, with emissions of 33 million tonnes of CO_2; in Poland, Belchatow spewing 29 million tonnes of CO_2;

---

[7]   Drax Group plc: Annual report and accounts 2013, www.drax.com/media/32649/drax_ar13_final.pdf

in the US, Scherer releasing 23 million tonnes of CO_2.[8] Mega numbers that are very hard to put into context.

So what does one tonne of CO_2 actually look like? It is actually about the same physical size as your average family semi-detached home – so 36 million tonnes of it is pretty huge.

The electricity flows out of these power stations and into the transmission system, the national grid, where we lose about another 9% transporting it around the country to your home, school or business. When it gets there we put it into lightbulbs, computers etc. that predominantly make heat instead of doing the task that they were designed for so there is yet more wastage. It turns out that only one in five units of energy actually do the thing we want them to do: light up our lives. The rest gets lost along the way.

This is the system we think of as modern, advanced even. This is the system that we are encouraging others to adopt across the world. To me that seems pretty mad. A system with nearly 80% of the energy being lost in the process surely can't be at the pinnacle of its development?

Flick the switch and consider the impact of what you are doing multiplied back upstream. Consider that five times the energy was used and five times

Waste in the current centralised energy system

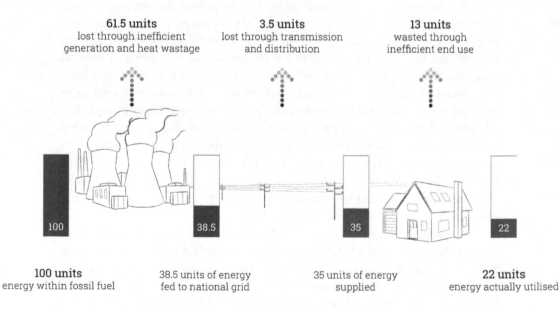

61.5 units lost through inefficient generation and heat wastage	3.5 units lost through transmission and distribution	13 units wasted through inefficient end use

| 100 units energy within fossil fuel | 38.5 units of energy fed to national grid | 35 units of energy supplied | 22 units energy actually utilised |

8   CAMRA – Carbon Monitoring for Action, http://camra.org

the pollution came out. If you are wondering why your bills are so expensive, massively wasteful system may be one of the reasons. This is a system that needs millions of tonnes of input each year, millions of acres of opencast mining and spews millions of tonnes of noxious gases.

The flip side is quite a contrast. Whilst those of us with access to this system take it for granted, there are still 1.5 billion people in the world with no access to electricity at all. The bloated system we have is mirrored in the extreme by many people around the world who have no access to energy. For three billion people access to energy still means wood for cooking, and that's it. Access to light, water and energy for cooking is pretty fundamental stuff and so many of our fellow humans don't have it.

Now, I am aware that I have talked a lot in this chapter about electricity production, when this only represents a fraction of our energy consumption, and heat is often the 'elephant in the room' when it comes to energy. Lots of our heat needs globally are met by natural gas, and this system has a different profile. Turn on your heating, and the gas ignites, flowing down pipes that link all the way to oil rigs out in the sea to gas terminals that receive liquefied gas shipped halfway round the world. There are losses along the way, compressing the gas to fit into tankers and shipping it around the world, keeping it cool so it stays compressed. But they are small in comparison to the losses on the electrical system. However, losses occur at our end of the pipe, in our buildings, our homes, and with our boilers. A huge proportion of the heat that we pump into our homes to keep them warm leaves the building as quickly as it has arrived! Lots of it is lost in inefficient boilers, whilst 25% of the heat flies up through the roof, 35% leaves through the walls – the list goes on. So we have a different challenge to address, another system which when you look at the whole of it also has a very low efficiency.

Transport is a similar story. Delivering oil around the world whilst incurring less losses along the way becomes very wasteful when we stick it in an engine that doesn't burn too efficiently and is wrapped in one and a half tonnes of metal that goes into making our cars.

In the developed nations, our lifestyles are like giant resource hoovers, sucking in stuff from all over the planet. We may have heard about this with food, but it is also true of energy. To give you a flavour, in the UK our plutonium comes from Australia and Canada, our gas from Qatar, Norway, Trinidad, Nigeria and Egypt. Our coal comes from Poland, Russia, the US, and Columbia. We do get some energy from our own country but at least 40% of what we use to make electricity is imported, as is an increasing amount of our oil and gas. This is a story that is repeated across the developed world, where very few developed nations are able to produce the energy they need at home. The US imported 40% of its oil requirements in 2012. China imports oil, gas and coal, and is in fact the largest producer and importer of coal in the world.

The dependence on other countries' resources to maintain our inefficient lifestyles is growing more extreme. If you look at oil, the predominant fuel used for transport, some of the world's biggest producers are Saudi Arabia, Iraq, Nigeria, Iran and Kuwait. All of them are famous for the conflicts, the interventions and the very strict regimes. Our needs for these resources put pressure on people around the world, and the history of troubles in one area of Nigeria is a testament to this. In the Niger delta there is a long history of oil spills, on a scale that makes the Gulf Horizon spill seem like a small event. There are pipelines running through farms and estuaries leaking their toxic contents and destroying people's homes and livelihoods, be that from sabotage or failure it is not totally clear. These spills have been going on for a long time and the impacts are huge. Fifty years of leaks are estimated to equal an Exxon Valdez spill every year.[9] Yet there is little international outrage or a huge clean-up response. The area is riddled with unrest and problems due almost certainly to the presence of 'black gold'.

You would perhaps expect that those of us that are benefitting from these giant companies would at least be grateful for their work. But that is far from the truth. Numerous surveys show that the utilities are not at the top of our lists for companies that we trust. In fact they are some of the most disliked institutions in our societies across all products and services.[10] Far from its entrepreneurial beginnings the utility companies across Europe are monsters born from nationalised industries, built over decades with billions of public investment. Privatisation occurred across Europe in the early 1990s, selling off this vast system to the highest bidder. This was intended to bring competition to a monopoly, but it didn't take long for the market to consolidate into a few very large companies, mirroring the monopoly, but privately owned, and spanning Europe entirely. A vision of the 'free market'!

So next time you flick the switch, or turn your ignition, think of the effects upstream, the pollution, and the impact on people's lives around the world. There must be a better way to do this!

---

9   'Far From Gulf, a Spill Scourge 5 Decades Old', www.nytimes.com/2010/06/17/world/africa/17nigeria.html?_r=0

10   'Big Energy Suppliers Slump to New Lows in Satisfaction Ratings', www.independent.co.uk/news/business/news/big-energy-suppliers-slump-to-new-lows-in-satisfaction-ratings-9070800.html

CHAPTER 3

FUNNY WEATHER

Super Typhoon Haiyan made landfall in my family's hometown and the devastation is staggering. I struggle to find words even for the images that we see from the news coverage. I struggle to find words to describe how I feel about the losses and damages we have suffered from this cataclysm. Up to this hour, I agonize while waiting for word as to the fate of my very own relatives. What gives me renewed strength and great relief was when my brother succeeded in communicating with us that he has survived the onslaught. In the last two days, he has been gathering bodies of the dead with his own two hands. He is hungry and weary as food supplies find it difficult to arrive in the hardest hit areas.[11]

Yeb Sano was lead negotiator for the Philippines in the Warsaw Climate Summit; he added this personal note in a speech that he made at the conference and it became a focus for the world's media. He did this because during the summit a super typhoon hit his country – the strongest tropical cyclone ever recorded with winds gusting up to 235mph. It left a huge trail of destruction, affecting 13 million people and leaving over 5,000 dead.[12] As the storm broke land – which it did four times over the Philippines – it brought with it storm surges and waves in some cases 17ft high. With waves big enough to flatten buildings, many areas were completely devastated with homes destroyed, infrastructure flattened and power and water cut. Three million people were displaced, 250,000 homes were destroyed and a further 250,000 were damaged.[13] It is unimaginable for those of us who have been fortunate to not live through an event like this to understand what is it like for so many people to have their homes and livelihoods destroyed in one go. As if the loss of life and injury were not enough, the legacy of a storm like this is huge, with a crisis of epic proportions to be faced and a country to be rebuilt afterwards.

This super typhoon is one of many extreme weather events that have happened all around the world. Powerful hurricanes in the US, huge flooding

---

[11] '"It's Time to Stop This Madness" – Phillipines Plea at UN Climate Talks', www.rtcc.org/2013/11/11/its-time-to-stop-this-madness-philippines-plea-at-un-climate-talks

[12] 'Typhoon Haiyan Death Toll Rises Over 5,000', www.bbc.co.uk/news/world-asia-25051606

[13] 'Typhoon Haiyan: A Crisis by the Numbers', www.wfla.com/story/23991224/typhoon-haiyan-a-crisis-by-the-numbers

events in Europe, heat waves, droughts and wildfires in Australia, record rainfalls and landslides in Brazil. These events – which have touched people on every continent – are an indicator that we are already living with a climate that has changed. These changes are already affecting more people each year than war and conflict – a recent study by the Norwegian Refugee Council said that in 2013, 22 million people were driven from their homes by floods, hurricanes and other hazards. Apparently that is twice the number affected in 1970, and many of these events go unreported.[14]

In fact since the 1950s the world has warmed by about 0.7°C according to the Intergovernmental Panel on Climate Change (IPCC). This may not sound like a lot, but with the rate at which we are emitting carbon dioxide we are on course to warm our world by 4°C, when compared to pre-industrial levels, by the end of this century. The last time the earth had an average temperature 2°C higher was 115,000 years ago, and global sea levels were 5m higher than they are now.[15] That would mean pretty much every major city in the world would be under water. This is not something we can plan for or come up with an easy solution for.

In the summary of their latest report The IPCC reports that:

Warming of the climate system is unequivocal, and since the 1950s, many of the observed changes are unprecedented over decades to millennia. The atmosphere and ocean have warmed, the amounts of snow and ice have diminished, sea level has risen, and the concentrations of greenhouse gases have increased. Each of the last three decades has been successively warmer at the Earth's surface than any preceding decade since 1850 .... In the Northern Hemisphere, 1983-2012 was *likely* the warmest 30-year period of the last 1,400 years.[16]

The thousands of scientists involved in writing the IPCC reports do so on an entirely voluntary basis and are collaborating across the world to reach consensus that is then released as their report.

But for many people the media coverage of both the science of climate change and the impacts are still very confusing. Despite the overwhelming scientific consensus on the subject, climate change is routinely reported with a question mark over whether it is actually happening or not. Of the 11,000 scientific papers written on the subject, over 97% conclude it is happening and

[14] 'Natural Disasters Displaced More People Than War in 2013, Study Finds', www.theguardian.com/world/2014/sep/17/natural-disasters-refugee-people-war-2013-study

[15] 'Climate Change is Here Now and it Could Lead to Global Conflict', www.theguardian.com/environment/2014/feb/13/storms-floods-climate-change-upon-us-lord-stern

[16] 'Summary for Policymakers'. In *Climate Change 2013: The Physical Science Basis*. Contribution of Working Group I to the Fifth Assessment Report of The Intergovernmental Panel on Climate Change [Stocker, T.F., D. Qin, G.-K. Plattner, M. Tignor, S.K. Allen, J. Boschung, A. Nauels, Y. Xia, V. Bex and P.M. Midgley (eds.)]. 2013, Cambridge University Press.

that humans are causing it.[17]

Yeb Sano summed it up pretty well in his speech:

> To anyone who continues to deny the reality that is climate change, I dare you to get off your ivory tower and away from the comfort of your armchair. I dare you to go to the islands of the Pacific, the islands of the Caribbean and the islands of the Indian Ocean and see the impacts of rising sea levels; to the mountainous regions of the Himalayas and the Andes to see communities confronting glacial floods, to the Arctic where communities grapple with the fast dwindling polar ice caps, to the large deltas of the Mekong, the Ganges, the Amazon, and the Nile where lives and livelihoods are drowned, to the hills of Central America that confronts similar monstrous hurricanes, to the vast savannahs of Africa where climate change has likewise become a matter of life and death as food and water becomes scarce. Not to forget the massive hurricanes in the Gulf of Mexico and the eastern seaboard of North America. And if that is not enough, you may want to pay a visit to the Philippines right now.

There still seems, however, to be a huge question in the media as to whether climate change is actually happening. Why this question is still being asked is interesting in itself. It mirrors the struggles that occurred in exposing the facts around the health issues linked to tobacco use. Some of the organisations and individuals that were at the heart of the tobacco health issue denial campaigns have popped up again with a similar approach for the climate issue. In fact there is a complex web of foundations, funds and experts in the US and the UK that all work systematically to discredit the climate science: "the aim of the campaign was to 'reposition global warming as theory rather than fact'".[18] They have names like the 'Global Warming Policy Foundation' and the 'Global Climate Coalition', and they have taken a range of tactics to keep a question mark over the certainty of climate science in the media. By questioning the science and releasing their own spurious 'scientific papers', by attacking individual scientists in the IPCC and trying to discredit the whole IPCC process. The agenda is simply to delay action, by seeding confusion and doubt in the non-expert, be they politician or voter. The various foundations and funds leading these attacks often do not disclose the sources of their funds and who their backers are[19] even though some of them are well known like the infamous Koch brothers. What is clear is that a whole lot of money is being

---

[17] 'Quantifying the Consensus on Anthropogenic Global Warming in the Scientific Literature', http://iopscience.iop.org/1748-9326/8/2/024024/article

[18] *Dealing in Doubt: The Climate Denial Machine vs Climate Science*. 2013, Greenpeace USA.

[19] 'Secret Funding Helped Build Vast Network of Climate Denial Thinktanks', www.theguardian.com/environment/2013/feb/14/funding-climate-change-denial-thinktanks-network
'Secret Funding of Climate Sceptics is Not Restricted to the US', www.theguardian.com/environment/2013/feb/15/secret-funding-climate-sceptics-not-restricted-us

put into these foundations every year. One recent study found that the annual average income of 91 of these US based 'climate change counter-movement' organisations was $900 million.[20] The US population have been left confused. A survey in 2012 asked, "Do scientists believe the earth is getting warmer because of human activity?" 43% responded no, and 12% didn't know.[21] Sadly it seems in the case of our elected leaders this disinformation has worked, in the UK one study said, "The vast majority of Tory MPs don't think manmade climate change is a real issue."[22] Whether this is a real representation is not clear – perhaps many MPs don't want to speak up about these issues because they don't understand the science fully, and actually don't want to become a target of the climate change deniers.

Despite this, some governments and NGOs have been trying to address this issue, and 1992 saw the world's first climate summit. The Rio Earth Summit as it was known resulted in the UN Framework Convention on Climate Change, where pretty much every country agreed that we should do something about climate change – they just didn't agree what. In subsequent years they have been meeting to try and pin down that 'what' into a tangible agreement. There has been some progress over the years, notably in Kyoto in 1997 and then in Copenhagen in 2009. However the emissions cuts signed up to by the wealthiest nations in 1997 were only followed through by Europe, with Russia and Canada backing out and the US never ratifying the treaty. In 2009 in Copenhagen, the world's nations agreed on how to define 'dangerous' global warming. Basically they agreed that we should limit global average temperature rises to no more than 2°C, and some of the largest polluters such as the US, China and India pledged to make voluntary cuts to their emissions. However these cuts won't keep us under the 2°C limit.

The annual UN climate conferences have seemed to the outsider to be lots of talk and not much action. As Yeb Sano put it: "This process under the UNFCCC has been called many names. It has been called a farce. It has been called an annual carbon-intensive gathering of useless frequent flyers." It can seem depressing when our leaders seem powerless to make meaningful progress on such an important issue, and it seems frustrating after all these years of discussions that there seems to be very little progress. In fact carbon emissions have continued to grow.

As I write this the latest series of climate talks are just about to take place, in the run-up to attempting to get a new agreement in 2015. Around the world 600,000 people have just taken to the streets in 2,000 locations around the

[20]   Brulle, Robert (2013). *Institutionalizing Delay: Foundation Funding and the U.S. Climate Change Counter-movement Organizations.* Springer Science and Business Media Dordrecht.

[21]   Pew Research Center, www.people-press.org/2012/10/15/more-say-there-is-solid-evidence-of-global-warming

[22]   'Tory Views On Climate Change Are A "Huge Embarrassment" For David Cameron', www.huffingtonpost.co.uk/2014/09/10/climate-change-tory-mps-david-cameron_n_5796978.html

world calling on our leaders for change,[23] but it is likely any agreement struck now will not come into force until 2020. That timetable will do nothing to help the people living in one of the many island states spread across our world. In the middle of the Pacific Ocean lies a nation of 33 islands, Kiribati. The 103,000 citizens of Kiribati may well become the first climate refugees. Their Prime Minister Anote Tong says of their situation:

> If nothing is done, Kiribati will go down into the ocean. By about 2030 we start disappearing. Our existence will come to an end in stages. First, the freshwater lens will be destroyed. The breadfruit trees, the taro, the salt-water is going to kill them. So we won't be able to maintain the integrity of all the islands. There's no high ground. So we will have to evacuate islands. We will defend the islands that we can, but we can't protect against storms. We have very moderate weather here, but if that changes with the climate, we won't survive. We would not survive a Hurricane Sandy. We would be finished. It would push the ocean across our islands.

The ancestors of the Kiribati have lived in harmony on the islands for at least 3,000 years, and they are now faced with losing their lands, their homes and their nation. The government of Kiribati recently bought 6,000 acres of land in Fiji, but do not have an invitation to settle there or anywhere else.[24]

As if that were not bad enough, one report from 2012 suggests that climate change is already contributing to the deaths of 400,000 people globally each year and costing $1.2 trillion, about 1.6% of GDP.[25]

Climate change is such a huge issue it is easy to feel very powerless in the face of it. Yeb Sano's response at the Warsaw summit surely highlights the despair and frustration of those trying to make changes happen on these issues: "In solidarity with my countrymen who are struggling to find food back home and with my brother who has not had food for the last three days, I will now commence a voluntary fasting for the climate. This means I will voluntarily refrain from eating food until a meaningful outcome is in sight (from the summit)." Many others have joined Yeb, and this has led to the monthly 'Fast for the Climate'.[26]

These issues seem huge, unsolvable and what is worse, we know we are complicit in them. To be put off from action at this point by those who say it will never make a difference is not an option, but that is the path most of us take. To be caught in the fear and despair are a real possibility for many of us, as these issues seem so beyond our control. But in the end our society is made

---

[23] 'Climate Change Summit: Global Rallies Demand Action', www.bbc.co.uk/news/science-environment-29301969

[24] www.businessweek.com/articles/2013-11-21/kiribati-climate-change-destroys-pacific-island-nation

[25] 'Climate Change is Already Damaging Global Economy, Report Finds', www.theguardian.com/environment/2012/sep/26/climate-change-damaging-global-economy

[26] Fast for the Climate, http://fastfortheclimate.org/en/

up of people like you and me; it is the choices we make that affect everything. Whilst our actions may seem insignificant, we have nothing else, and as the saying goes, 'If you think small things don't make a difference try spending the night with a mosquito.'

> *If the world is to be healed through human efforts, I am convinced it will be by ordinary people, people whose love for this life is even greater than their fear. People who can open to the web of life that called us into being.*
>
> Joanna Macy, author, environmental activist
> and teacher of The Work That Reconnects

Me too. It is time to move beyond the fear that many of us feel deep down about these issues. Stepping into the sadness and grief, to a point where action based in positivity once more becomes possible. Bringing this positivity and inspiration that a different future is possible to your community and your town – probably the most compelling reason to take action. We cannot expect our leaders to do this alone; it is a subject that needs us all to lead or at least be engaged with finding and applying the solutions. Even with the positive moves of some countries, like Denmark pledging to be 'fossil free' and Sweden announcing it was aiming for 'net zero' emissions by 2050 at the recent summit,[27] we must play our part.

For me, playing my part started by putting my body in the way of bull-dozers on opencast mine and motorway construction sites in the early 1990s. An extreme action that many people considered pretty mad at the time, and probably still would. What it achieved in the grand scheme of things is probably not a lot. But for me personally, I now understand, it was the first step in overcoming the sadness, despair, grief and anger and transforming them into action. Once engaged in action things started to change for me to a point where what I wanted to do was build solutions. That is what I have spent my life focused on since then.

We face the biggest challenge of our species, the threat to this spaceship on which we are travelling. It is something that we need to achieve in the coming few decades, taking measurable steps along the way. The only option is to transform our society to repair the damage, and that needs us to tap into the creative source to dream up solutions. It needs all who can to get engaged and play our part. As Joanna Macy puts it:

> The biggest gift you can give is to be absolutely present, and when you're worrying about whether you're hopeful or hopeless or pessimistic or optimistic, who cares? The main thing is that you're showing up, that you're here and that you're finding ever more capacity to love this world because

---

[27] '10 Best Leaders' Pledges at the UN Summit', www.rtcc.org/2014/09/24/10-best-leaders-pledges-at-the-un-summit

it will not be healed without that. That was what is going to unleash our intelligence and our ingenuity and our solidarity for the healing of our world.

It always seems impossible until it's done.

Nelson Mandela

It is time to make the impossible possible.

CHAPTER 4

FRACKING SALVATION

We are told that oil, coal and gas are the stuff that our civilisation was made by, the key drivers of our economic engine, and in many ways they have been. The last century was certainly the century of coal, oil and gas. But if we carry on with the thinking and technology that got us here, it is also apparent that we won't have a planet left that is habitable for humans and many other species in the immediate future. Now oil, coal and gas have become a destructive and wasteful addiction that we really need to give up for the sake of all life. You may think that the use of the word 'addiction' is a bit strong, but when you look at our relationship with fossil fuels, we can see that it displays many of the classic symptoms of addictive behaviour. Our use continues despite awareness of the problems, and to maintain a good supply we are taking more and more risks. All over the world social and environmental problems have been caused by our desire to extract these substances. Our need for the stuff has become so extreme that we are destroying vast tracts of land and leaving huge toxic waste pools in the process. We are drilling wells beneath our homes and through our water supplies and poisoning the air and water that we and all life depend on. While I don't really want to dwell on the destructive techniques now being employed, I feel compelled to give you just a taste of the madness that is being carried out in our names. If we continue to ignore these things, we will struggle to change them.

Many people have written on the subject of 'peak oil', but what really hits me is seeing the wholesale destruction of beautiful areas of the world to harvest oil for us to drive our cars, and gas to heat our homes or make electricity. Oil extraction and gas production have often been out of sight for many of us, in Texas, Saudi Arabia or offshore. It is easy for us to forget the huge supply chain behind the nozzle pumping petrol into our cars and piping gas into our homes. But increasingly, the supply chain is getting closer to our homes and communities, with threats to our land and water. We hear of it occasionally, when there is a spill or accident, or when wars are fought in oil-rich areas, as has been happening pretty often in the last few decades.

The oil and gas that is easy to get at has basically been used up. Most of the traditional wells and resources have reached a point where they are declining in production, and to make up the difference we are having to get oil and gas from less ideal and more costly and damaging places. The traditional oil fields in the US have been in decline for decades, producing 10 million barrels a day in the 1970s versus 6 million today.[28] The production of crude oil in the UK peaked in around 1999, pumping some 4.5 million barrels per day.[29] But, quietly in the background, extraction has been evolving; as prices have gone up, it has opened up all sorts of new avenues.

The melting ice in the Arctic seas mean that Shell and other oil giants have begun operations to try and exploit the potential reserves under the sea bed in those regions[30] – with headlines like 'Arctic drilling is inevitable: if we don't find oil in the ice, then Russia will' – just to reinforce the seeming inevitability of the situation. The prizes may be big, but the risks are also huge, with the potential of oil spills causing major damage to the ecology of the area, one of the last wildernesses on earth – not to mention the effect on the global climate. The prize is estimated to be 166 billion barrels of oil – enough to provide current consumption of oil for about five years, so actually not that huge. But to get the prize, they will be working in extreme conditions in the middle of a watery wilderness. The test drilling carried out by Shell in 2012 is perhaps a portent for the challenges faced in this environment, as Shell's drilling rig, *Kulluk*, ran aground off the coast of Alaska. The US Coast Guard said the incident was due to the company's 'inadequate assessment and management of risks' in icy, storm-tossed waters.[31] It seems that they had been trying to move the rig to avoid millions of dollars in tax liability when it broke free of its towing vessel in high seas. Luckily, the rig lodged itself on a rock shelf, with no loss of life or spillage of the many tonnes of diesel it had on board. Drilling for oil in such a remote location is risky, especially considering the challenges that followed the explosion at the Deepwater Horizon platform which took months to stop, killed 11 people in the initial explosion, and oiled hundreds of miles of coastline. That rig was relatively close to the teams needed to get it under control; one in the Arctic would not be.

Although efforts to extract oil from the Arctic have so far been unsuccessful, there are some other extremes happening much closer to our homes to fuel our profligate lifestyles. People living in Alberta, Canada, are in the middle of the front line of fossil fuel extraction – oil sand. With the easy-flowing crude oil hard to come by, attention has turned to the shales and sands that contain bitumen that can be squeezed out of them. The technique has been around for 40 years, but it has only gained traction recently, as oil prices increase to a level

---

[28]   'Has the United States Beaten Peak Oil? Not so Fast', www.washingtonpost.com/blogs/wonkblog/post/has-the-united-states-beaten-peak-oil-not-so-fast/2012/02/17/gIQAhFbAKR_blog.html

[29]   'North Sea Oil: Facts and Figures', www.bbc.co.uk/news/uk-scotland-scotland-politics-26326117

[30]   'Arctic Drilling is Inevitable: If We Don't Find Oil in the Ice, Then Russia Will', www.telegraph.co.uk/finance/newsbysector/energy/11080635/Arctic-drilling-is-inevitable-if-we-dont-find-oil-in-the-ice-then-Russia-will.html

[31]   'Coast Guard Blames Shell Risk-Taking in Kulluk Rig Accident', http://news.nationalgeographic.com/news/energy/2014/04/140404-coast-guard-blames-shell-in-kulluk-rig-accident/

that makes it economically viable. They may be called 'sands', but the bitumen sets hard so it is more like rock than beach sand. The process is messy. Millions of tonnes of these rocks are moved by truck to processing plants where they are heated to release the oil from the sand. When I say 'millions of tonnes', that is actually an understatement: they are in fact moving a million tonnes of the stuff each day. The scale of this operation is hard to believe. In many cases a hundred feet of earth has to be stripped off to reach the deposits. About two tonnes of bitumen-laden sand are required to produce a single barrel of oil. The three-story high trucks carrying the sand haul 400-tonne loads, burning 50 gallons of diesel an hour in the process. The rocks are sticky and need to be crushed to start with, and are then mixed with water – about 200,000 tonnes of water is being used each day – which is heated to wash out the bitumen. The bitumen then needs to be heated and pressurised to make it useable.[32]

Sadly, the process creates a lot of waste water that is laced with a range of toxic chemicals. Some of this waste water is re-used in the process, but most remains untreated in what are known as 'tailings lakes'. These tailings lakes are 30-50% of the mines' footprint and, to give you an idea of the scale of the issue, by 2010 the lakes covered 176 square kilometres of land and contained approximately 830 million cubic metres of toxic water. All this water is drawn from the Athabasca River, and there are concerns that at certain times of low water flow, the rates of extraction may cause damage to the river's ecology. For each barrel produced, 1.5 barrels of tailings waste are added to the landscape, that's 1.5 barrels of toxic water. These tailings contain a host of toxins including bitumen, naphthenic acids, cyanide, phenols and metals such as arsenic, cadmium, chromium, copper, lead and zinc. The tailings lakes seep approximately 11 million litres per day, risking contamination of surface water and groundwater systems. The lakes stay ice free all year round, because they are fed with warm water used in the process and therefore attract wild birds and other animals, which inevitably are harmed or even killed in the process. There have been a number of incidents of flocks of bird landing on them and subsequently dying.[33] To make this crazy situation even worse, there isn't really a plan of what to do with the tailings afterwards, just talk of putting them down a mineshaft with a 'cap' of freshwater on top. Many square miles of land are left stripped and in need of restoration – the miners are required only to reclaim the land, rather than restore it to its original state. By 2011 there were approved applications to mine 28,000 hectares of peat lands. Of the more than 70,000 hectares that had already been mined that year, only a tiny fraction have been reclaimed, and there is no way to return them to the old-growth boreal forest that once dwelled there.

---

[32] 'The Canadian Oil Boom', http://ngm.nationalgeographic.com/2009/03/canadian-oil-sands/kunzig-text/4

[33] Flanagan, Erin and Grant, Jennifer (2011). *Losing Ground: Why the Problem of Oilsands Tailings Waste Keeps Growing*. Pembina Institute.

For the reserves that are too deep for strip mining there is another technique used to get at the bitumen in situ. Wells are drilled into the seam of oil sands and steam is pumped down there to melt the bitumen so that it can be pumped out to the surface to undergo similar processing. While this process takes less land area and doesn't leave vast areas degraded, it still needs well pads, roads, power lines and other infrastructure running into the forest, and of course creates waste. The Canadian boreal forest covers two million square miles, of which around 75% remains undeveloped. Oil sand developments remain a tiny fraction of that land area, but if the *in situ* extraction technique is developed widely, this could change.

This whole process affects not only the land but also, of course, the people and other creatures touched by these developments. In February 2009, the Alberta Cancer Board released a study responding to community calls for further investigation. While the report determined that the number of cases of cholangiocarcinoma was within the expected range, the report found that the overall cancer rate was approximately 30% higher than expected. In response, Alberta Health Services issued a press release indicating that there was no problem. But it was revealed that there was a 30% increase in cancers in Fort Chipewyan, a community downstream of the processing plant, compared with expected rates over the last 12 years. There was also a threefold increase in leukemia and lymphomas, and a sevenfold increase in bile duct cancers. According to Natural Resources Defence Council Senior Scientist Dr. Gina Solomon:

> Leukemia and lymphomas have been linked in the scientific literature to petroleum products, including VOCs (volatile components of petroleum), dioxin-like chemicals and other hydrocarbons. Biliary cancers have been linked to petroleum and to polycyclic aromatic hydrocarbons (PAHs, chemicals in tar and soot). Soft tissue sarcomas are very rare and lethal cancers that have also been linked to dioxin-like chemicals and hydrocarbons. It's an interesting pattern – almost all of the cancer types that were elevated have been linked scientifically to chemicals in oil or tar.[34]

Tragic.

It's not just the people's health that is threatened but also an entire way of life. A large proportion of the people living in the area are from the First Nations. The caribou herds living in the area affected by oil sands are important for cultural and spiritual reasons, and they have declined 71% since 1996.

---

[34] Droitsch, Danielle and Simieritsch, Terra (2010). *Canadian Aboriginal Concerns With Oil Sands.* Pembina Institute.

> *The extinction of caribou would mean the extinction of our people. The caribou*
> *is our sacred animal; it is a measure of our way of life. When the caribou are*
> *dying, the land is dying.*
>
> Chief Janvier, Chipewyan Prairie Dene First Nation

The First Nations have expressed their concerns and called for a moratorium on all oil sand developments. They have tried to halt the progress of oil sand developments in the courts. But oil sands are big business. Approximately 390,000 Canadian jobs were linked to the oil sands in 2010, and the government of Alberta collected $4.5 billion in royalties from oil sand production, which represents 38.7% of non-renewable resource revenue and 11.4% of total government revenue.[35] The US imports more oil from Canada than any other country.[36]

Elsewhere in Canada, the US and Europe, another fossil fuel boom has been making the headlines. The delightfully named 'fracking' (hydraulic fracturing) technique has spread like wildfire across some areas of the US and is attempting to spread across areas of Europe and the rest of the world as well. Fracking has some high-level supporters, as well as a large grassroots movement in opposition. It has been hailed as the ultimate in 'low carbon' energy security by some, and a danger to health and environment by others. Its use is increasingly opposed across the world.

Fracking is an adaptation of the original oil-drilling process, for cases where the gas or oil doesn't naturally flow out of the seam because it is trapped in 'shale' rocks. The process involves pumping a high-volume mixture of water, sand and chemicals into the shale under pressure. The wells themselves don't just go straight down, but are drilled horizontally through the shale seam, running for miles underground and fanning out from the well location. Water and fracking fluids are then pumped into the well and, at extreme pressures, cause fractures in the rocks to release the gas trapped there. More than a million wells have been drilled and fracked in the US,[37] while in France, Germany and Bulgaria it has been banned due to environmental concerns. In the UK, only a handful of shale gas wells have been drilled despite cheerleading from Prime Minister David Cameron: "We cannot afford to miss out on fracking."[38]

Fracking for shale gas has been a big deal in the US, with those one million wells producing a lot of gas and changing the reliance of the US on imports. According to the US Energy Information Administration, total US net imports of energy as a share of energy consumption fell to their lowest level in 29 years

[35] Lemphers, Nathan and Woynillowicz, Dan (2012). *In the Shadow of the Boom: How Oilsands Development is Reshaping Canada's Economy*. Pembina Institute.

[36] www.eia.gov/dnav/pet/pet_move_impcus_a2_nus_ep00_im0_mbbl_m.htm

[37] www.api.org/oil-and-natural-gas-overview/exploration-and-production/hydraulic-fracturing/hydraulic-fracturing-qa.aspx

[38] www.ft.com/cms/s/0/a4e24b70-35ac-11e3-b539-00144feab7de.html#axzz3FbGkKes4

for the first six months of 2014.[39] In fact US production of crude oil surpassed all other countries this year with daily output exceeding 11 million barrels in the first half of the year. The US became the world's largest producer of natural gas in 2010.[40] Shale oil production now provides up to 20% of US consumption, with shale gas providing up to 40%. According to one report in 2012 this energy boom added almost $75 billion in federal and state revenues, contributed $283 billion to the gross domestic product and supported 2.1 million jobs.[41]

But there is a flip side to this equation. A report from 2010 called 'Fractured Communities' documents many cases of negative effects on local communities and local environments caused by the relentless push for fracking.

> In the past two years in Pennsylvania, state regulators have found that gas drilling using high-volume hydraulic fracturing has contaminated drinking water, polluted surface waters, polluted air, and contaminated soils. In Ohio, state regulators found that inadequate well casing resulted in drinking water contamination and the explosion of a house. In Texas, state regulators found elevated levels of benzene and other toxics in neighbourhoods with nearby gas compressors. In Wyoming, EPA has warned residents not to drink the water, and in Colorado, hundreds of spills have been reported as residents continue to investigate localized health impacts they feel are associated with nearby drilling operations.[42]

This wasn't just one or two incidents, but many, across hundreds of sites, over extended periods. Gas migration from wells seems to be a common problem, where gas starts to come into the water sources, rivers, streams and even buildings in areas near a well site. The pressure generated in the fracking process forces the fracking fluid – water laced with sometimes toxic chemicals and methane – into other fissures in the rocks, allowing it to rush back up to the surface. Of course this isn't meant to happen; wells have a casing that should make it impossible. But sometimes these casings fail or there are other paths for the gas-laden water to travel along. In one case in Pennsylvania, a landowner noticed that his well – which had a normal water level about 80 feet below the ground – was overflowing. On surveying his property he discovered around 20 geysers on his property gushing water infused with methane out of the ground. The basement of his house filled with methane and the fire service

[39]   'Net Energy Imports as Share of Consumption at Lowest Level in 29 Years', www.eia.gov/todayinenergy/detail.cfm?id=18351

[40]   'U.S. Seen as Biggest Oil Producer After Overtaking Saudi Arabia', www.bloomberg.com/news/2014-07-04/u-s-seen-as-biggest-oil-producer-after-overtaking-saudi.html

[41]   'Fracking Boom Seen Raising Household Incomes by $1,200', www.bloomberg.com/news/2013-09-04/fracking-boom-seen-raising-household-incomes-by-1-200.html

[42]   Michaels, Craig, Simpson, James and Wegner, William (2010). *Fractured Communities: Case Studies of the Environmental Impacts of Industrial Gas Drilling*. Riverkeeper, Inc.

asked him to evacuate because of the risk. Eventually Chesapeake, which had drilled the well, installed a ventilation pipe and filtration system to try and make his water useable, but methane was still found bubbling up in puddles all over this land.[43] This experience seems to be repeated fairly often, leaving some communities not trusting their water supplies or the fracking companies that are working in their area.

Another thing that has created suspicion over the fracking companies' activities is their refusal to disclose what chemicals they are mixing with the water they are using for fracking, as well as the sheer volumes of water being used in the process. According to an April 2011 report for the US House of Representatives Committee on Energy and Commerce, oil and gas companies use 750 chemicals during fracking. Many of the substances used in the process are fairly innocuous, but many are not. Naphthalene, xylene, toluene, ethylbenzene and formaldehyde, for example, each used in a number of proprietary fracking solutions, are known or suspected human carcinogens.[44] While the quantities of these chemicals in use seem small in percentage terms – perhaps 0.05-2% of the total volume – each well might need between 2-4 million gallons of water per frack, so many tonnes of the chemicals may be used. In most cases the water required for the frack arrives by truck, placing a huge strain on the local infrastructure and adding to the stress on local communities. Spillages of the 'produced water' that comes out of the well at the end of the process have also occurred in hundreds of cases. This produced water contains not only the fracking chemicals, but salts, heavy metals and some radionuclides drawn deep from within the earth. Not nice stuff. The spills find their way into local watercourses, resulting in damage to the aquatic ecology and fish deaths.

The list goes on. The produced water often sits in open pits, without any actual treatment. There have been quite a number of cases of well blow-outs and fires. The drilling disturbs local residents with vibration and constant low-level noise. The smells and fumes that come from the sites can cause irritation for local people. But probably most worrying are the clusters of human health problems that seem to appear around the well sites. The tricky thing with the health problems around these sites is that they are easy to dispute, and, as with climate change, there are counter studies claiming that there are no problems. However, reading some of the reports interviewing people about their health problems and concerns for their animals is disturbing. In one report from 2012 entitled 'Impacts of gas drilling on human and animal health', Michelle Bamberger and Robert Oswald describe how they spent a year speaking to

[43] 'In Northeast Pennsylvania, Methane Migration Means Flammable Puddles And 30-Foot Geysers', http://stateimpact.npr.org/pennsylvania/2012/07/30/in-northeast-pennsylvania-methane-migration-means-flammable-puddles-and-30-foot-geysers/

[44] 'Blind Rush? Shale Gas Boom Proceeds Amid Human Health Questions', www.ncbi.nlm.nih.gov/pmc/articles/PMC3237379/

people living in the shadow of the wells. None of the subjects in the report are identified by name, since many of the cases were in litigation at the time of writing. They describe quite a number of cases in which hundreds of farm animals suddenly died when exposed to fracking fluids, where leaks or spills have occurred. They also describe cases of stillborn livestock in those animals that survived. The people who cared for these animals also had health problems such as extreme fatigue, headaches, nosebleeds, rashes and sensory deficits (smell and hearing).

"Soon after drilling and hydraulic fracturing began for the first well, a child began showing signs of fatigue, severe abdominal pain, sore throat, and back-ache. Six months later, the child was hospitalized with confusion and delirium and was given morphine for abdominal pain."

It turned out that these symptoms were a sign of arsenic poisoning – a direct result of the fracking and, on the advice of the doctor, the family moved away to allow the child to recover.[45] One study on birth outcomes for mothers living close to fracking wells concluded that there was a "positive association between greater density and proximity of natural gas wells within a 10-mile radius of maternal residence and greater prevalence" of neural tube defects (NTDs) and congenital heart defects.[46] Chilling stuff. Both these reports said that more studies are needed, and that effective study is hampered by the gagging orders and non-disclosure agreements often signed by the people affected by these terrible circumstances. Is it worth these risks? Is this the future we want for our children?

As if all that were not enough to make people question the technique, many parts of America are facing the most severe drought for years and fracking is adding to the problem. A report by investor network Ceres stated: "Nearly half of the wells hydraulically fractured since 2011 were in regions with high or extremely high water stress, and over 55% were in areas experiencing drought."[47] So, in some regions, fracking and food growing are competing for water, a very precarious situation.

In North Dakota they have been drilling for shale oil. A by-product is the production of gas. However, the infrastructure required to affordably get much of this gas to market is lacking, so the gas is simply burnt, generating an intense light that can be seen at night from space. If you look at the pictures, you would think that a new mega city had been born, but no, it is 'flared' wasted gas lighting up the night sky. "Every day, drillers in the Bakken burn off about 350 million cubic feet of natural gas. That comes to more than $100

[45]   Bamberger, Michelle and Oswald, Robert (2012). 'Impacts of Gas Drilling on Human and Animal Health'. In *New Solutions* Vol. 22(1) 51-77.

[46]   'Birth Outcomes and Maternal Residential Proximity to Natural Gas Development in Rural Colorado', Lisa M. McKenzie, Ruixin Guo, Roxana Z. Witter, David A. Savitz, Lee S. Newman, and John L. Adgate, http://ehp.niehs.nih.gov/1306722/

[47]   Freyman, Monika (2014). *Hydraulic Fracturing & Water Stress: Water Demand by the Numbers – Shareholder, Lender & Operator Guide to Water Sourcing*. Ceres.

million worth of gas burned off each month."[48] Regulators are stepping in to stop this, but it is hard to believe that it happens at all.

Yet with all the above, we are still being bombarded with messages that fracking and shale gas offer us an opportunity for low bills, energy security and an important 'transition fuel' on the journey to a low-carbon future. This last point may be true, in simple terms, displacing output from a coal-fired power station with gas-powered generation, but when you factor in gas flaring and methane migration, the position is not so clear-cut. Assumptions that gas will replace coal can also not be guaranteed. It is more likely that increased gas use would simply mean that renewable energy isn't deployed as fast – so the coal keeps being burned at the current rate. According to David Cameron, 'fracking has real potential to drive energy bills down'. His piece in the *Telegraph* newspaper is nothing short of a top line sales pitch, he goes on, "If we don't back this technology, we will miss a massive opportunity to help families with their bills."[49] Contrasting this view, his Energy Secretary Ed Davey said of its effect on gas prices, "We can't expect UK shale production alone to have any effect."[50] Even Lord Browne, former BP chairman, advisor to David Cameron and former chairman of fracking company Cuadrilla said, "We are part of a well-connected European gas market and, unless it is a gigantic amount of gas, it is not going to have material impact on price."[51] Unlike the US, the gas can simply be piped around Europe and sold to the highest bidder.

What is behind the rhetoric from on high about fracking? It turns out that since 2010, Hillary Clinton has been promoting the technique all over the world, supporting the major oil companies in their search for fracking sites in new countries, and creating a new bureau to lead the charge.[52] Playing catch-up four years on, the UK Prime Minister went on record saying, "Fracking will be good for our country". He blamed some of the opposition to shale gas on a 'lack of understanding' of the process.[53] However, people across the UK disagreed, and a new movement of people willing to put themselves in the way of the trucks was born at the Sussex village of Balcombe. At the peak of the protest many people were arrested, as well as the only UK Green Member of Parliament, Caroline Lucas.[54] The protests continue everywhere the fracking

48 'Inside North Dakota's Latest Fracking Problem', www.cnbc.com/id/101934384#
49 'We Cannot afford to Miss Out on Shale Gas', www.telegraph.co.uk/news/politics/10236664/We-cannot-afford-to-miss-out-on-shale-gas.html
50 '"Fracking Won't Lower Energy Bills", says Davey', www.telegraph.co.uk/earth/energy/fracking/10296274/Fracking-wont-lower-energy-bills-says-Davey.html
51 'Lord Browne: Fracking Will Not Reduce UK Gas Prices', www.theguardian.com/environment/2013/nov/29/browne-fracking-not-reduce-uk-gas-prices-shale-energy-bills
52 'How Hillary Clinton's State Department Sold Fracking to the World', www.motherjones.com/environment/2014/09/hillary-clinton-fracking-shale-state-department-chevron
53 'Fracking "Good for the UK", Says David Cameron', www.theguardian.com/environment/2014/mar/26/fracking-good-for-uk-cameron
54 'Caroline Lucas Among Dozens Arrested in Balcombe Anti-fracking Protest', www.theguardian.com/environment/2013/aug/19/caroline-lucas-arrest-balcombe-anti-fracking

crews show up, yet the government keeps pushing the agenda and trying to move the obstacles out of the way, changing the laws on trespass and the environmental legislation to make it easier for fracking to take place – despite promising the opposite.

We may reasonably suppose from all this that there is a lot of money in shale and that is why it is being so aggressively pushed, but it seems there is even a question mark over that. The growth of production from shale has been spectacular, but it is also common to find a very steep rate of decline from the initial peak of production. This has led many commentators to liken the growth of the shale business to the dotcom era. The basic issue seems to be that, because of the steep decline rate of the wells, to keep production stable you have to keep drilling new ones, and that costs a lot of money. As the *Telegraph* puts it:

> Shale has been hyped ('Saudi America') and investors have poured hundreds of billions of dollars into the shale sector. If a huge number of wells come on stream in a short time, you get a lot of initial production. Output falls away very quickly indeed after production begins. Compared with 'normal' oil and gas wells, where output typically decreases by 7-10% annually, it is by no means unusual for production from each well to fall by 60% or more in the first 12 months of operations alone. The only way to keep production rates up (and to keep investors on side) is to drill yet more wells.[55]

Bloomberg went on:

> Drillers are caught in a bind. They must keep borrowing to pay for exploration needed to offset the steep production declines typical of shale wells. At the same time, investors have been pushing companies to cut back. Spending tumbled at 26 of the 61 firms examined. For companies that can't afford to keep drilling, less oil coming out means less money coming in, accelerating the financial tailspin.[56]

I think this is enough to give you a pretty good picture of the current state of madness that is our energy system, but if you want more, go and look up coal bed methane or mountaintop removal to complete the picture. These involve burning coal seams while they lie underground to get a gas from them, or demolishing whole mountains of coal, all so we can keep our houses warm and our cars moving!

As a young man fresh out of university where I studied climate science,

[55] 'Shale Gas: "The Dotcom Bubble of Our Times"', www.telegraph.co.uk/finance/personalfinance/investing/11006723/fracking-for-Shale-gas-the-dotcom-bubble-of-our-times.html

[56] 'Shakeout Threatens Shale Patch as Frackers Go for Broke', www.bloomberg.com/news/2014-05-26/shakeout-threatens-shale-patch-as-frackers-go-for-broke.html

as well as energy engineering, I was pretty angry about the situation and decided to try and do something to stop the destruction. For me, that culminated in a valley in South Wales, where the ex-coal miners invited me and my friends to come and help them defend the last few acres of old-growth woodland and water meadow in the area. The rest had been opencast mined, and then 'restored'. These areas were like a moonscape. Trees, many planted years before as part of the 'restoration', were stunted and still looked as though they were only a few years old. The only parts left were the village and these few acres. The coal miners had lost their jobs and the opencast companies had moved in. We built treehouses and rope walkways in an attempt to protect the trees, and lay in front of the diggers and trucks to try and stop their progress. But, despite our best efforts, with the help of many policemen and security guards, the last oaks in that valley fell.

At some point we will have to stop the destruction economy before our lands and waters, as well as our climate, are irreversibly damaged. The extreme lengths we are currently going to in order to meet our energy needs are surely the wakeup call that there must be a better way? Even those in the oil industry themselves know the end is in sight. Art Berman, a Houston-based geological consultant, put it like this: "I'm all for shale plays, but let's be honest about things, after all. Production from shale is not a revolution; it's a retirement party."[57]

Let's hope he is right.

Perhaps a portent of things to come, the very town where fracking started has just won a battle to get the practice banned within the city limits. Sharon Wilson, the Texas organiser for EarthWorks said of the achievement: "It should send a signal to industry that if the people in Texas – where fracking was invented – can't live with it, nobody can."[58] To top it off, New York Governor Andrew Cuomo announced he will ban hydraulic fracturing, stating, "I will be bound by what the experts say." His officials said the potential health and environmental impacts were too great to allow fracking to proceed in the state at the time; they also highlighted studies regarding the long-term safety of hydraulic fracturing. "I think it's our responsibility to develop an alternative … for safe, clean economic development," the Governor said. Quite right.

---

[57] 'Is the U.S. Shale Boom Going Bust?', www.bloombergview.com/articles/2014-04-22/is-the-u-s-shale-boom-going-bust

[58] 'Texas Oil Town Makes History as Residents Say No to Fracking', www.theguardian.com/environment/2014/nov/05/birthplace-fracking-boom-votes-ban-denton-texas

CHAPTER 5

NUCLEAR RENAISSANCE

Nuclear power is often hailed as a part of the solution to our energy and climate problems. A few years ago I came across a speech by a UK minister promoting the future use of nuclear energy, and to me it just seemed like a catalogue of all the reasons why you would never want to build new nuclear power stations.

"Nuclear policy is a runner to be the most expensive failure of post-war British policy-making," said Chris Huhne, the then Secretary of State for Energy and Climate Change, in a speech to the Royal Society on 13th October 2011. Strong words from a Secretary of State.

He continued in the same speech:

We currently have around 6,900 cubic metres of high-level nuclear waste. That's about enough to fill three Olympic swimming pools. We have enough intermediate-level waste to fill a supertanker, and a lot more low-level waste.

We manage the world's largest plutonium stocks – more than a hundred tonnes – and they will need guarding for as long as it takes us to convert it and build long-term deep storage. And if we don't, we will have to guard it for tens of thousands of years.

Half of my department's budget goes in cleaning up this mess, and it will rise to two thirds next year. That is £2 billion a year, year in and year out that we are continuing to pay for electricity that was consumed in the fifties, sixties and seventies on a false prospectus.

Yet the total nuclear liabilities that the Nuclear Decommissioning Authority now deal with are estimated to be £49 billion, and I cannot be confident that the figure will not rise again as we discover yet more problems.

Just look at the history of rose-tinted spectacles: the provisions for nuclear decommissioning costs in total were £2 million in 1970, £472 million in 1980, £9.5 billion in 1990, £22.5 billion in 2000. And now £53.7 billion.[59]

---

[59] 'The Rt Hon Chris Huhne MP Speech to the Royal Society: Why the Future of Nuclear Power Will Be Different', www.gov.uk/government/speeches/the-rt-hon-chris-huhne-mp-speech-to-the-royal-society-why-the-future-of-nuclear-power-will-be-different

By 2014, just three years on, the estimated bill in today's money is £64.9 billion, with an actual cost of £110 billion. The complexity of the task at hand means that there is huge uncertainty as to what the actual final cost will be. The Nuclear Decommissioning Authority of the UK said in June 2014, "Potential costs could be somewhere between £88 billion and £212 billion."[60]

All of this expenditure is effectively paying for the electricity our parents had over the last four decades, and is creating no actual power.

ENERGY TOO CHEAP TO METER

The nuclear industry actually set out with noble aims and lofty ambitions. As Lewis Strauss, then Chairman of the United States Atomic Energy Commission, said in a 1954 speech to the National Association of Science Writers:

> Our children will enjoy in their homes electrical energy too cheap to meter ... It is not too much to expect that our children will know of great periodic regional famines in the world only as matters of history, will travel effortlessly over the seas and under them and through the air with a minimum of danger and at great speeds, and will experience a lifespan far longer than ours, as disease yields and man comes to understand what causes him to age.

Nuclear power has been sold to the public at different times and under very different pretences. In 1956 the world's first reactor providing power that was supplying to the electricity grid went into operation in the UK. Billed under the same terms of providing 'energy too cheap to meter' it was actually constructed to create plutonium for nuclear weapons.

But the theme of 'energy too cheap to meter' seems to have stuck with the technology, even though we can see from the UK clean-up figures alone that it is far from that.

NEW NUCLEAR

If there is to be a nuclear renaissance we will need to build a whole load of new reactors – but if they were truly such a cost-effective and easy form of power generation you would expect that energy corporations would be queuing up to build them. Well, at the moment there are 66 nuclear reactors in construction globally, two thirds of which are in Russia, China and India. Of these 66, nine have been under construction for more than 20 years and at least 23 of them have experienced significant construction delays.

---

[60] *Insight into Nuclear Decommissioning Including Financial Highlights – 14th edition.* 2014, Nuclear Decommissioning Authority. p.13. www.nda.gov.uk/publication/insight-into-nuclear-decommissioning-including-financial-highlights-14th-edition-june-2014/

The thing to remember is that, according to the most recent report on the state of the nuclear industry, it is actually in decline globally:

> The 427 operating reactors are 17 lower than the peak in 2002, while, the total installed capacity peaked in 2010 at 375GWe before declining to the current level, which was last seen a decade ago. Annual nuclear electricity generation reached a maximum in 2006 at 2,660TWh, then dropped to 2,346TWh in 2012 (down 7% compared to 2011, down 12% from 2006). About three-quarters of this decline is due to the situation in Japan.[61]

Currently there are two new nuclear power stations in construction in Europe. One, the Olkiluoto 3 EPR in Finland, was originally planned to go online early in 2009, and it now seems that it will not be producing electricity before 2018. The 1.6GW reactor was originally priced at €3 billion, ramped to €8.5 billion in 2012, and the cost estimates are still rising alarmingly. So much for a fixed turnkey price! At the moment the Finnish operator and the French constructor are in contractual dispute with one party claiming compensation of €1 billion for alleged failures, and the other demanding €2.4 billion in compensation for delays. Meanwhile in France, the state owned nuclear corporation EDF have confirmed that their new build reactor at Flamanville is experiencing significant time and cost overruns. Originally scheduled to start operating in 2012, EDF now 'hope' that the reactor may be operational by 2016. Originally priced at €3.3 billion, the bill for the completed reactor is currently estimated at €8.5 billion.

So building these complex machines is getting more expensive than ever, and taking much longer to complete, with the average new construction taking 9.4 years.[62]

Where new construction is planned, it is common for governments to be involved, particularly in providing loan guarantees for construction costs. In the US the federal government provided loans and loan guarantees of $8.3 billion for the construction of a new reactor in Georgia.[63] The UK government meanwhile has agreed to a loan guarantee and loan of £17 billion to try to cope with the inevitable construction risk for the proposed new nuclear power station at Hinkley Point C by the French state owned EDF.[64]

Once constructed, new nuclear power stations still require government subsidies to make them financially viable. In the UK, this is supposed to come from a new mechanism called a 'Contract for Difference'. In the case of Hinkley Point C, a price of £92.50 will be paid per mega watt hour (MWhr)

[61] Froggatt, Antony and Schneider, Mycle (2013). *World Nuclear Industry Status Report*. p.6.

[62] Froggatt, Antony and Schneider, Mycle (2013). *World Nuclear Industry Status Report*. p.7.

[63] 'U.S. Loan Guarantee Set for Georgia Nuclear Plant', http://online.wsj.com/news/articles/SB100014 24052702304275304579393500420131562

[64] 'U.K. Nuclear Plan Advances With $15 Billion Loan Backing', www.bloomberg.com/news/2013-06-27/u-k-s-nuclear-plan-advances-with-15-billion-treasury-backing.html

– approximately twice the price of wholesale electricity today.[65] This payment will be index linked and will last for 35 years, more than twice the length of any similar subsidies given to renewable generators. This may well add up to a £100 billion payout over the life of the plant, flowing from UK electricity consumers and taxpayers to EDF.

WASTE – THE MILLION YEAR QUESTION

We have already heard the scale of the waste situation in the UK and this is of course similar in all countries around the world with nuclear power stations. Radioactive wastes are not simple things to deal with, and the estimated time frame for dealing with radioactive waste ranges from 10,000-1,000,000 years.[66] Humanity could be a different species by then!

At the moment, the entire financial burden for cleaning up the existing nuclear power stations is borne by taxpayers. In the case of EDF's proposed new power station at Hinkley Point C, they have struck a deal which caps their liability for decommissioning, with any cost overruns being picked up by the UK taxpayers once more. Smart move.

Decommissioning a nuclear power station is a complex and costly thing that can take a very long time. According to the UK Nuclear Decommissioning Authority, this is a task that is estimated to take 100 years. The waste then needs to be stored for a very, very long period of time. In the UK, proposals for nuclear waste storage involve building a geological disposal facility (GDF) – basically a very deep hole in the ground. Except no one currently knows how much one of these will cost, and no one has successfully constructed or operated one of these anywhere in the world. Unsurprisingly not many communities are very keen on the idea of siting one near them.

The list of operational problems with new and existing nuclear goes on. As sea levels rise, more money will need to be spent on keeping the reactors, and their interim waste stores, dry. In fact this has already happened with one UK power station, Dungeness, being taken offline whilst extra work was undertaken on the flood defences of the site which 'were not as robust as previously thought'.[67] Inland power stations are also suffering from the effects of climate change. In France in 2009, 14 of the 19 nuclear power stations located on rivers were taken offline due to overheating issues.[68]

---

[65] 'UK Agrees Nuclear Power Deal with EDF', www.ft.com/cms/s/0/00eff456-3979-11e3-a3a4-00144feab7de.html#axzz3CukDYXod

[66] *Technical Bases for Yucca Mountain Standards*. National Research Council. 1995, National Academy Press.

[67] 'Exclusive: Dungeness Nuclear Power Station Quietly Taken Offline for Five Months Over Fears of Fukushima-style Flood Disaster', www.independent.co.uk/news/uk/home-news/exclusive-dungeness-nuclear-power-station-quietly-taken-offline-for-five-months-over-fears-of-fukushimastyle-flood-disaster-9200494.html

[68] 'France Imports UK Electricity As Plants Shut'. *The Times*. 3rd July 2009.

LOW-CARBON ELECTRICITY

The big question is: Will nuclear power help us with managing global warming by helping to reduce our CO_2 emissions? It is talked about as a 'zero carbon' technology. However it does push out significant CO_2 emissions through uranium mining and milling, transport, fuel enrichment, plant construction, operation, plant decommissioning and waste management. Whilst the reported range of emissions for nuclear energy over the lifetime of a plant is from 1.4g of carbon dioxide equivalent per kWh (g CO_2e/kWh) to 288g CO_2e/kWh, the mean value is 66g CO_2e/kWh.[69] Although emission values are still lower than those of coal or oil (600-1,200g/kWhel), they remain significantly higher than for wind (2.8-7.4g/kWhel), hydropower (17-22g/kWhel), photovoltaic (19-59g/kWhel), and energy efficiency measures (which are *circa* 10 times more cost effective).[70]

So it looks like it really isn't the 'magic bullet' in terms of climate emissions.

WORST-CASE SCENARIO

Unfortunately nuclear is most famous for the accidents. Chernobyl, Fukushima, Three Mile Island and Windscale are the well-known ones but there have been many smaller incidents that go unmentioned.[71] It seems there is a culture of secrecy that surrounds nuclear energy – perhaps a hangover from the days when it was also producing weapons grade plutonium, but very damaging nonetheless. A year after its opening, on 10th October 1957, there was a serious accident at the Windscale power station in the UK. It was the world's first nuclear disaster, the facts of which were not made public for 30 years.

The most recent accident at Fukushima Daiichi in Japan, on 11th March 2011, has left a huge legacy of waste and pollution. Around 150,000 people have been forced to evacuate the area, and may not be able to return to their homes and farms for many years – if ever. The reactors remain in a fragile condition, requiring many tonnes of water to be pumped into them daily to keep them cool. The radioactive water coming out is stored in makeshift tanks, and transported in temporary piping, that tend to leak. By May 2013, 380,000 tonnes of irradiated water had been used for cooling, and the volume is increasing daily. Water is also seeping from tanks underneath the reactor into the sea and so far the only plan to try and stop this is to install a wall of ice in the ground to prevent the out-flow. Contamination is already showing

---

[69] Sovacool B.K. (2008). 'Valuing the Greenhouse Gas Emissions from Nuclear Power: A Critical Survey'. In *Energy Policy* Vol. 36, pp.2940-2953.

[70] Wallner A., Wenisch A., Baumann M., Renner S. (2011). *Energy Balance of Nuclear Power Generation, Life-cycle Analysis of Nuclear Power: Energy Balance and CO₂ Emissions*. Österreichisches Ökologie-Institut, Austrian Climate and Energy Fund, Vienna.

[71] 'Revealed: Catalogue of Atomic Leaks', www.theguardian.com/environment/2009/jun/21/nuclear-power-stations-inspector-watchdog

up in the fish, molluscs, shellfish and seaweed – which were a huge part of the economy of that area. The long-term nature of this problem cannot be underestimated with TEPCO, the Japanese nuclear utility, saying it may take 40 years to retrieve the fuel from the melted reactors – and in the meantime the whole site remains in a vulnerable and fragile condition.

> The worst-case scenario, as depicted by the Chairman of the Japan Atomic Energy Commission in the middle of the crisis in March 2011, remains the collapse of the spent fuel pool of unit 4 and a subsequent fuel fire, potentially requiring evacuation of up to 10 million people in a 250 kilometre radius of Fukushima, including a significant part of Tokyo.[72]

What is probably most surprising about the disaster at Fukushima is that three reports written on the subject, one by the Japanese government – one by an independent team of experts in Japan and a third by The Carnegie Endowment for International Peace — have all concluded that this disaster was not a 'natural' disaster after all.

"It was a profoundly man-made disaster that could and should have been foreseen and prevented," said Kiyoshi Kurokawa, chairman of the Fukushima Nuclear Accident Independent Investigation Commission.[73]

Estimated costs for the legacy of Fukushima are currently tens to hundreds of billions of euros, a monstrous sum. However in Europe, nuclear accident liability for any one accident is currently only capped at €169 million.[74] These figures are worlds apart! The fact that the share price of the world's largest nuclear operator, the French state utility EDF, fell by 85% over the past five years, seems to be the true barometer for the future of nuclear energy globally.

Nuclear power and its legacy of waste and pollution will be with us for many years. It is going to take many decades to even begin the task of remedying some of the current problems and dismantling the current reactors. There are still so many unanswered questions as to what we will do with the toxic and persistent wastes that need safe storage for millennia.

One would hope that with decades of public subsidy and support it would truly be a cost-effective form of energy, but that is far from the case. Adding new generation capacity will increase our collective exposure to the many risks involved. Radioactively contaminated land and waste, for which we have no real plan, is an unfair and dangerous legacy to leave our children. If we are really serious about climate change and energy, we should be focusing our efforts on dealing with our vast current liabilities rather than making more insoluble problems for future generations.

---

[72] Froggatt, Antony and Schneider, Mycle (2013). *World Nuclear Industry Status Report*. p.9.
[73] 'Disasters, Leadership and Rebuilding – Tough Lessons from Japan and the U.S., Institute for Global Environmental Leadership,' http://environment.wharton.upenn.edu
[74] 'Hinkley C – A Nuclear Subsidy Too Far', www.theecologist.org/News/news_analysis/2347944/ hinkley_c_a_nuclear_subsidy_too_far.html

CHAPTER 6

SUBSIDY JUNKIES AND FOOL'S GOLD

It is like running both the heating and the air conditioning at the same time.

Former BP chairman Lord Browne on fossil fuel subsidies[75]

Perhaps like me you would expect that such a developed market as the fossil fuel industry would have weaned itself off public financial support. You may be surprised to hear that that is far from the case. The International Energy Agency estimates that globally, the cost of government subsidies for fossil fuels increased from $311 billion in 2009 to $544 billion in 2012. The IMF concluded that the actual cost when factoring in lost government revenues is actually around $2 trillion, around 8% of all government revenues.[76] The IEA goes on to conclude that without further reform, spending on fossil fuel consumption subsidies is set to reach $660 billion in 2020, or 0.7% of global GDP.[77] These are massive sums for industries that are so well established, and so harmful to our climate. In 2013 the IEA's Chief Economist Dr Fatih Birol described fossil fuel subsidies as 'public enemy number one' for the production of sustainable energy. He went on to say this represented an incentive to emit carbon equivalent to $110 per tonne. In contrast, the EU emissions trading system currently provides a disincentive to emit carbon of less than $10 per tonne.[78] Duncan Clark of The Guardian sums it up pretty well in his piece on energy subsidies back in 2012: "On a planet staring devastating climate change in the face, spending taxpayers' money on propping up fossil fuels really is as crazy as throwing buckets of petrol on a house fire."[79] Much of this subsidy is in developing nations, reducing the prices of oil to stimulate economic growth.

---

[75] 'Lord Browne: Fracking Will Not Reduce UK Gas Prices', www.theguardian.com/environment/2013/nov/29/browne-fracking-not-reduce-uk-gas-prices-shale-energy-bills

[76] 'Fuelling Controversy', www.economist.com/news/finance-and-economics/21593484-economic-case-scrapping-fossil-fuel-subsidies-getting-stronger-fuelling

[77] 'IEA Analysis of Fossil-fuel Subsidies', www.iea.org/media/weowebsite/energysubsidies/ff_subsidies_slides.pdf

[78] 'IEA chief: 'Fossil Fuel Subsidies are Public Enemy Number One for Green Energy', www.businessgreen.com/bg/news/2241226/iea-chief-fossil-fuel-subsidies-are-public-enemy-number-one-for-green-energy

[79] 'Fossil Fuel Subsidies: A Tour of the Data', www.theguardian.com/environment/datablog/2012/jan/18/fossil-fuel-subsidy

However there are questions as to whether this subsidy actually helps the people who need it or just makes strong multinationals stronger still. I think you can guess the answer. The IEA estimates that $22.5 billion was spent by India on fossil fuel subsidies in 2010 but that less than $2 billion benefitted the poorest 20% of the population.[80]

Meanwhile renewable energy is constantly plagued with accusations of only being viable because of public subsidies. In comparison global subsidies for renewable energy in 2012 were $101 billion.[81] Germany is often held up as the country that has burdened its economy by investing heavily in renewable energy, using public subsidy to drive that growth. However between 1979 and 2010 subsidies in Germany were as follows: Coal and lignite received €222 billion, nuclear €186 billion and renewable energy received €28 billion.[82] Globally less than one fifth of the totals for fossil fuels in the same year, and the myth of the free market continues. It is true that renewables currently generate a small amount of the energy when compared to fossil fuels so in terms of subsidy taken right now they are more costly. However they are at the start of their deployment so the subsidy being given is to help get the technologies to scale. Unlike fossil fuels that have been subsidised for many years, and continue to be.

There is a darker shadow looming around the economics of fossil fuels. The Carbon Tracker[83] initiative has been doing pioneering research into the amount of CO_2 we can release to stay within two degrees of warming that climate scientists deem prudent and that governments have accepted in international agreements. It turns out the known reserves of fossil fuels if burnt would actually represent emissions of three times this amount of carbon we can safely emit. As Mark Carney, the governor of the Bank of England, recently put it, the "vast majority of reserves are unburnable".[84] President Obama's climate envoy, Todd Stern, has said the same thing: "It is going to have to be a solution that leaves a lot of fossil fuel assets in the ground."[85] These known reserves go some way to informing the valuation of the companies involved. These companies make up a huge section of our economy, and the value of stock markets. If their reserves can't be burned, that entirely changes their valuations and has the potential to create the next big financial crash, potentially making sub-prime look like a warm up. As chair of the House of Commons Environmental Audit

[80]   'If Fossil Fuel Subsidies Are so Bad, Why Can't We Get Rid of Them? Time for Some Politics', http://oxfamblogs.org/fp2p/if-fossil-fuel-subsidies-are-so-bad-why-cant-we-get-rid-of-them-time-for-some-politics/
[81]   'Fossil Fuel Subsidies Growing Despite Concerns', www.bbc.co.uk/news/business-27142377
[82]   'Global Cooling – Strategies for Climate Protection', p72, Hans-Josef Fell
[83]   www.carbontracker.org
[84]   'Mark Carney: Most Fossil Fuel Reserves Can't Be Burned', www.theguardian.com/environment/2014/oct/13/mark-carney-fossil-fuel-reserves-burned-carbon-bubble
[85]   'Obama's Climate Change Envoy: Fossil Fuels Will Have to Stay in the Ground', www.theguardian.com/environment/2014/nov/25/todd-stern-fossil-fuels-ground-climate-change-obama

Committee (EAC) Joan Walley MP put it when launching a report on the issue: "Financial stability could be threatened if shares in fossil fuel companies turn out to be overvalued because the bulk of their oil, coal and gas reserves cannot be burnt without further destabilising the climate."

Many of Europe's pension funds hold fossil fuel company stock and Green MEP Reinhard Butikofer explained the risk as follows: "With over €1 trillion in high-carbon assets, we have identified that the carbon bubble is a significant risk, particularly for a number of EU Member States and EU financial institutions. Investments in fossil fuel companies could therefore quickly turn into fool's gold."[86]

The oil majors seem unconcerned, but have been pushed to respond to people's concerns. Shell responded in a letter to shareholders:

> ... because of the long-lived nature of the infrastructure and many assets in the energy system, any transformation will inevitably take decades. This is in addition to the growth on energy demand that will likely continue until mid-century, and possibly beyond. The world will continue to need oil and gas for many decades to come, supporting both demand, and oil and gas prices. As such, we do not believe that any of our proven reserves will become stranded.[87]

So not only does fossil fuel extraction present a huge threat to our life support system – the climate – it presents a huge potential economic risk, if we are serious about addressing the first problem. Right now Shell's response is simply business as usual works for us thanks very much.

More and more investors are starting to wake up to the problem and in recent months there has been a rash of high profile 'divestments' where funds have started to move the money they manage out of fossil fuel investments. In fact globally there is a campaign afoot to create a 'divestment' movement, and it is gaining traction. It is a technique that has been tried and proven in previous struggles with corporate bad practice like with the tobacco industry. Recent fossil fuel divestments include the University of Glasgow, the British Medical Association, the World Council of Churches, Stanford University and the Rockefeller Foundation. The Church of Sweden was one of the first institutions to take up divestment. Gunnela Hahn, Head of Responsible Investment at the Church of Sweden said:

> As a responsible investor we look upon ourselves as owners of the companies we invest in. We do not want to own, and thereby fund, the extraction

---

[86] '"Carbon Bubble" Threatens Stock Markets, Say MPs', www.bbc.co.uk/news/science-environment-26455763

[87] 'What the Fossil Fuel Industry Thinks of the "Carbon Bubble"', www.carbonbrief.org/blog/2014/07/we-ask-the-fossil-fuel-industry-what-it-thinks-of-the-carbon-bubble

of fossil fuels. Instead we want to own and fund companies that stand for solutions. Furthermore we see a financial risk in owning fossil fuel companies. Their value consists to a large extent of fossil fuel reserves that risk losing in value, since they cannot be extracted if we are to have a liveable planet.[88]

The impact of moving money from these companies is probably insignificant in itself in these early stages. However according to a report by the Stranded Assets Programme at University of Oxford's Smith School of Enterprise and the Environment,[89] divestment "creates far more indirect impact by raising public awareness, stigmatising target companies and influencing government officials". It goes on to say this is likely to create new market norms, "increase legislative uncertainty and potentially also lead to multiple compression causing more permanent damage to the companies' enterprise values".

So it is the start of a new chapter for fossil fuel companies, where not only are their jobs getting harder, the value of their assets are going to be getting smaller. But right now the tables are still turned very much in their favour and if there was one thing we should be lobbying our governments for, it is to take away the subsidies that help keep the advantage further weighted on their side.

[88] 'Rapid Increase in Institutions Pulling Money Out of Fossil Fuels', http://gofossilfree.org/press-release/rapid-increase-in-institutions-pulling-money-out-of-fossil-fuels/
[89] 'Stranded Assets and the Fossil Fuel Divestment Campaign: What Does Divestment Mean for the Valuation of Fossil Fuel Assets?', www.smithschool.ox.ac.uk/research/stranded-assets/SAP-divestment-report-final.pdf

PART II

Stories from the Revolution

Our dependence on fossil fuels amounts to global pyromania, and the only fire extinguisher we have at our disposal is renewable energy.

HERMANN SCHEER[1]

---

[1] http://en.wikipedia.org/wiki/Hermann_Scheer

CHAPTER 7

THE ENERGY REVOLUTION

There is an energy revolution happening but no, it is not being televised.

Around the world people in many countries are working to deploy renewable energy technologies at a scale and pace that is transforming the way energy is used and generated. It is starting to disrupt the existing markets for energy.

There are so many stories out there that illustrate the speed of change in the deployment of clean energy technology. Yet amazingly, many times when I stand up in front of a group and say that there is a renewable revolution happening, people find it hard to believe. It seems the common misconception that has been put out is that renewables will never take the place of fossil fuel energy when quite the opposite is starting to happen. Renewable energy and particularly solar energy, is rapidly becoming the cheapest form of energy in many parts of the world and deployment rates are rocketing. But looking at the traditional media that is often the last impression you would get.

The energy revolution is truly the next great transition, the fourth revolution of human development. First there was the agricultural revolution, second the industrial revolution, third the information revolution and now the energy revolution is underway. It is essential that this happens, and I believe it is also inevitable. The price of fossil fuels has become extremely volatile, and extraction of new sources gets more complex and expensive. At the same time the effects of pollution and climate change are becoming more acute and clear for us all to see. It is time for energy generation and use to catch up with human development, it is time for energy generation to be bought back into sync with natural cycles. Most of the energy systems we use now were essentially invented in the industrial revolution and stemmed from the burning of coal as the engine of progress. It is time for the wholesale transformation of that system – as Amory Lovins so aptly puts it in his book of the same name – it is time for 'reinventing fire'. It seems that this transformation will not be led solely by governments or industry, but must be led by people in their local communities.

Many of us have lived through huge changes in the last two decades as the information revolution has transformed the way we communicate, shop, do business, are entertained, and work. Twenty years ago when people were buying the first personal computers, many people questioned why anyone would actually want a computer in their home. It seems hard to believe considering how deeply intertwined this technology has become in people's lives. The mobile telephone revolution has had an even greater impact and it has grown at a staggering pace. In 1990 there were hardly any mobile phones out there and yet today just over 20 years on there are over four billion of them in use worldwide. Now many of us take it for granted that the device in our pocket will be able to make a call, surf the net, play music and film as well as access social media platforms for the rapid sharing of ideas and information. This is a huge development in a short space of time.

This communications revolution has been disruptive in developed economies where there were landline infrastructures and established systems of working, but it has been totally transformative in Africa. It was estimated that there would be one billion mobile users on the continent by 2015, and the use of mobile phones has had some surprising positive effects. Social networks using mobile phones have increased the reporting of crises and in turn the response to them. They have increased the involvement in and safety of elections. Mobile phones are revolutionising financial services in Africa. In Kenya the company M-PESA is a true global success story in the field of mobile payments with over 18 million active users. People who had no access to banking can suddenly access it through this platform. This transforms what is possible for people who are often still living without electricity and running water. It transforms their ability to purchase things that would otherwise be unattainable, and it opens many new avenues. Ebooks are now accessible to millions of Africans and there are also mobile based services offering healthcare tips, maternity advice, tips for farmers and up to date weather reports. When so many of the population are smallholders dependent on what they can grow for livelihood, this can be essential information. The mobile and information revolution is enabling people around the world to do things differently for the first time.[1]

The energy revolution has the promise to be as far reaching as the previous revolutions and without a doubt needs to be as transformative. There are still 1.5 billion people on the planet with no access to electricity and three billion who still cook on smoky and dangerous wood, dung and charcoal fires. Those of us in the developed nations have tapped into a wealth of energy in the form of fossil fuels that have transformed our societies and lives. Access to energy that is a daily struggle for many has become an instant and careless process for many others. In developed nations we are running bloated and inefficient systems where in many cases huge percentages of the energy resource is simply wasted – straight up the chimney. So this energy revolution has two distinct paths to walk: A powering down or 'repowering' path for the wasteful and heavily energy dependent countries, and a powering up path for those for whom access to energy is still a daily struggle.

[1]   'How Africa's Mobile Revolution is Disrupting the Continent', http://edition.cnn.com/2014/01/24/business/davos-africa-mobile-explosion/

There are many inspiring stories of where both these paths are already being walked in many communities around the world, and in this section I will tell some of those stories. There are many inspirational people forging the new paths for our communities, and making the change in their own town. Some of them have shared their stories here.

The energy revolution will require a change not only in the way we capture, use and store energy, but also a shift in how we finance its construction and who owns the systems. It will inevitably move from a centrally owned and controlled system to one that is distributed and networked. We have already shifted to a distributed system when it comes to information – the World Wide Web has created a networked system of information storage. This system is dynamic and amazingly adaptable, with critical functions operating from multiple locations, so if there is a server failure in one place, there are others that will keep the system working. Our new energy system needs to be similar, encompassing a mix of technologies, optimised for local conditions, and delivering energy from a range of sources to provide a stable supply.

DECENTRALISED ENERGY SYSTEM

Once before we sourced all of the energy we needed for our lives from the land directly around us. Our primary sources for energy were wood, wind and water. For many people around the world this is still the case. In developed nations we have built up huge infrastructures to source the energy we need from all over the world, sucking resources from all over the planet for our needlessly wasteful lifestyles. Now it is time to wind the clock back and once again source all of our energy needs from the natural resources directly around us rather than from limited resources shipped around the world. It is also time to use those resources with care so that we make the most out of the energy we have available to us. Renewable energy technologies and energy efficiency techniques make this possible. What many people see as a pipe dream has already been achieved in some places – with technologies and systems that are readily available today.

Your roof can become a power station, your town could turn its waste food into biogas, its sewage waste could become a source of energy, your local river or tidal stream could generate power. We have a range of technologies, many already technically proven and commercially available that could enable this transition to a local energy production. There are many examples of communities who have embraced these ideas and made the transition, with many positive side effects. There are already towns that generate more energy than they need, and have created businesses and jobs in the process. There are countries that in moments have generated more energy than they need from renewable energy alone. The decentralised energy system is already under construction.

The decentralised energy town
How it might look

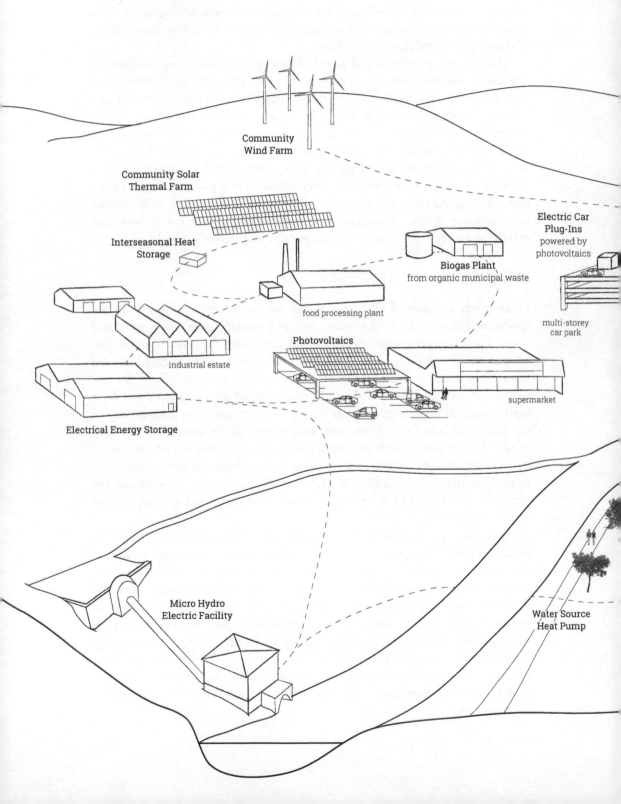

Community
Wind Farm

Community Solar
Thermal Farm

Electric Car
Plug-Ins
powered by
photovoltaics

Interseasonal Heat
Storage

Biogas Plant
from organic municipal waste

food processing plant

multi-storey
car park

Photovoltaics

industrial estate

supermarket

Electrical Energy Storage

Micro Hydro
Electric Facility

Water Source
Heat Pump

The decentralised energy system cuts out many of the losses associated with the current centralised system, transmitting power over vast distances from remote power stations or transporting fossil fuels by tanker from faraway places. It also offers a model for developing nations to adopt directly, without the need to build big, inefficient and expensive centralised infrastructure.

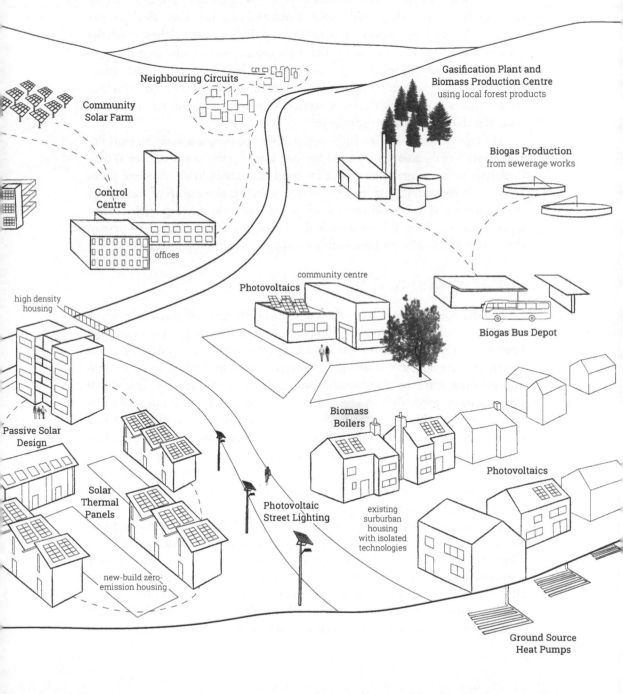

ENERGY FOR GOOD

Access to energy is fundamental to life. From cooking to reading to washing, so many essential acts in life require energy. Yet so many people do not have access to the energy they need for life and not just in developing nations.

Amazingly in developed nations where we all use massive amounts of energy every day there are still huge numbers of people who do not have access to the basic levels of energy they need to stay warm and run their lives. Fuel poverty is rife in the developed world. Here in England, in one of the richest countries on Earth in 2014, it is estimated that there are 2.2 million children living in fuel poverty – where their parents have to choose between heating and eating. So not only are we running massively wasteful and fossil fuel hungry systems, we are also not providing the basic services to many people in our societies, a situation that surely cannot continue.

The energy revolution is a huge opportunity for people across the world to take control of the energy they need for their lives, for the first time ever. It is an opportunity for energy to be a force for good rather than just a source of profit for large corporations. For that to become a reality, ownership of the energy generation and distribution assets needs to transfer from the corporations back to local communities. It is time we had a democratically owned energy system that met needs locally, and renewable energy gives us this opportunity.

ENERGY DEMOCRACY

One of the potential great side effects of mass deployment of renewable energy schemes is the shift in ownership that can occur in the energy system and has been widely reported in Germany. The nature of renewable energy developments is that they are built at a whole range of scales, from individual home systems and business sized, to community sized and very large systems. Traditional power stations generally only come in one size – very large. This range of scales that renewable schemes are built in means that investment is opened up to a much wider audience. It is now not just larger multinational corporations that can afford to build, own and operate a power station. You can, I can, your business can, your local council can, your local farm and community can. In many ways the small-scale modular nature of the technology makes it even harder for large corporations to get involved.

The analysis of ownership of all of the renewable energy systems deployed in Germany by 2010 is amazing – 62% was owned by individuals and farmers. Only 6.5% was owned by the large energy corporations. The large energy corporations rely on you and me being at the end of the wire, and the moment that you are not, the moment that you become a generator yourself, there is one less customer for them to rely on. This perhaps doesn't sound like a big deal, but when you look at the speed with which these changes occur you can understand the huge implications. In the UK between 2010 and 2013, we went from a having a few thousand solar PV powered homes to having over half a million of them. That's half a million people who have an insignificant or massively reduced electricity bill. There are around 25 million homes in the UK so these people taking control of their energy will make a significant impact on the traditional energy businesses. In Germany, of course, the numbers are much higher. Couple these solar homes with the energy storage technology that is on its way and the utility that provides your energy right now becomes a small player in your energy landscape.

This mass transference of ownership redesigns the system entirely. As we have seen earlier it fundamentally undermines the business models of the big energy corporations because you won't only be at the end of the pipe any more. You will be interacting with the grid – essentially using it as a storage and transfer mechanism. When you are no longer solely a consumer, the economics of the entire system shifts, and perhaps it only takes a shift of a few percentage points for things to change forever. Suddenly we need a new word to define your role in this new system – and the one I have heard in relation to this is 'prosumer'. You become a producer as much as a consumer – trading electricity or energy as you do with many other things in your life.

Renewables in the hands of the people
Ownership of renewables installed capacity in Germany, 2010

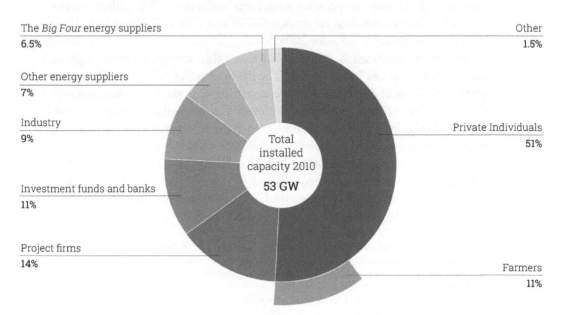

The *Big Four* energy suppliers
6.5%

Other
1.5%

Other energy suppliers
7%

Industry
9%

Private Individuals
51%

Investment funds and banks
11%

Project firms
14%

Total installed capacity 2010

53 GW

Farmers
11%

Source: German Energy Transition, energytransition.de, www.unendlich-viel-energie.de

THE END OF THE UTILITY?

With consumers becoming 'prosumers' and an influx of renewable energy sources coming onto traditional networks, the business models of the utilities that dominate European and US energy supply are starting to look outdated. In fact it seems that they are starting to be threatened by them, as the system changes from centralised fossil fuel and nuclear power to a distributed renewable one. There have been numerous write-ups of the phenomenon and the early effects in some countries. As summed up by Duke Energy CEO, Jim Rogers, "If the cost of solar panels keeps coming down, installation costs come down and if they combine solar with battery technology and a power management system, then we have someone just using [the grid] for backup."[2]

On some days in Germany, for instance, the grid is awash with effectively free solar and wind energy, so that the wholesale price for electricity falls to levels that are no longer profitable, in fact on some days the wholesale price has effectively gone negative. The middle of the day when solar output and consumer energy demand is at its highest, was normally the time when utilities made the most money. Now however, those peaks in power are often being provided by the 30GW of solar that has priority access to the grid. This has led to a reduced output from fossil fuelled power plants and a corresponding reduction in revenues as a consequence. German electric utility E.ON forecast revenues to dive 43% in 2013 compared to the previous year, according to Thomson Reuters. Rival utility RWE's Chief Executive Peter Terium, said in November 2013 that his operating model was 'collapsing'.[3] This reflects on the value of these companies as well; both E.ON and rival RWE have lost about 70% of their market value since 2008.[4]

Whilst the penetration of solar and renewable energy is not so high in many other countries in the world, it is having similar effects in other areas where high levels of deployment are being achieved. In the US' sunniest states, California and Arizona, the threat is already on the horizon. In response to the rise of solar in Arizona a tax has been introduced on customers with leased solar panels to try and slow the market down, it would seem.[5]

The end game here is that utilities that have invested in expensive fossil fuel and nuclear power station 'assets' will end up with two hits on their business models – a reduction in revenues, and a set of expensive assets that are no longer needed. The more expensive the assets – and the more costly to run –

2   'Solar Panels Could Destroy U.S. Utilities, According to U.S. Utilities', http://grist.org/climate-energy/solar-panels-could-destroy-u-s-utilities-according-to-u-s-utilities/
3   'RWE to Cut Jobs as Green Energy Expansion Hits Wholesale Prices', www.ft.com/cms/s/0/b769e958-4d06-11e3-9f40-00144feabdc0.html
4   'US Utilities Face German Style Solar Burn', www.breakingviews.com/us-utilities-face-german-style-solar-burn/21125829.article
5   'Solar Companies Sue Over New Rooftop Solar Tax In Arizona', http://thinkprogress.org/climate/2014/07/01/3455140/solar-companies-sue-arizona-tax/

the worse it will hurt, nuclear operators with high fixed costs stand to lose the most. It is likely that to compensate for the loss in revenues, the utilities will put up their prices and yet more of their customers will start to generate their own power from renewable as a result, compounding the situation.

In the US, utilities are faced with competition from new solar start-ups like Solarcity, which provides solar energy systems to homeowners and businesses. CEO Lyndon Rive says of the company: "We install solar systems for free, and we sell the electricity at a lower rate than you can buy it from the utility. So given the option of paying more for dirty power or paying less for clean power, what would you take?"[6]

Electricity production in Germany in week 30, 2015
by source – shows the huge impact of wind and solar on the German grid

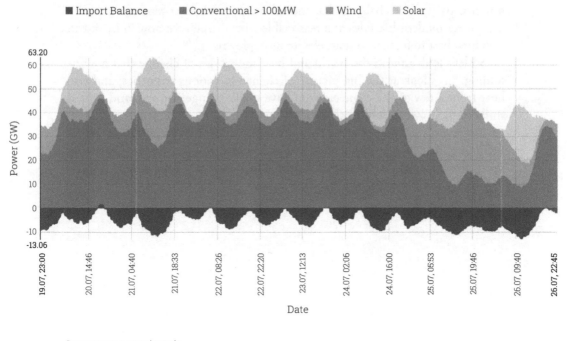

Source: www.energy-charts.de

---

[6]   'SolarCity CEO Talks the Future of Solar Power', http://fortune.com/2013/03/04/solarcity-ceo-talks-the-future-of-solar-power/

REPOWERING

Theories are marvellous things but personally I have always got more from hearing about actual examples of where the theories have been shown to work. There are already many amazing headlines that illustrate the success of renewable energy across the world. For instance, wind power was the largest single source of electricity production in Spain in 2013. Renewable energy accounted for 42.4% of the total electricity demand in Spain during 2013, 10.5% higher than in 2012.[7] It was estimated by Portugal's electricity network operator that renewable energy supplied 70% of total consumption of energy in the first quarter of 2013.[8] In 2012, the percentage of gross final consumption of energy that came from renewable sources in Sweden was 51%, in Latvia was 35.8%, in Finland was 34.3% and in Austria was 32.1%.[9]

According to the International Renewable Energy Agency, "Worldwide, renewable power capacity has grown 85% over the past 10 years, reaching 1,700GW in 2013, and renewables today constitute 30% of all installed power capacity. The challenge has moved on from whether renewable energy can power modern lifestyles at a reasonable cost – which we now know it can – to how best to finance and accelerate its deployment."[10]

So let's look around the world and hear the stories of change that are unfolding. Let's look at the big picture of change in various countries, and also hear from a few of the pioneers who have made change happen, and caused shifts in their own communities.

There are two themes at work – one of communities 'repowering' or re-designing their energy supply in developed nations, and the other of 'powering up' – providing access to energy for people that currently have none.

Get ready for the fun bit!

---

[7] 'Wind Power Was Spain's Top Source of Electricity in 2013', www.theguardian.com/environment/2014/jan/06/wind-power-spain-electricity-2013

[8] 'Is 70 Percent Renewable Power Possible? Portugal Just Did It For 3 Months', http://thinkprogress.org/climate/2013/04/14/1858811/is-70-renewable-power-possible-portugal-just-did-it-for-3-months/

[9] 'Share of Renewables in Energy Consumption up to 14% in 2012', www.unendlich-viel-energie.de/share-of-renewables-in-energy-consumption-up-to-14prozent-in-2012

[10] 'Rethinking Energy Towards A New Power System', www.irena.org/rethinking

CHAPTER 8

DENMARK

On 25th January 2014, 99% of the electricity consumed in Denmark came from wind turbines alone.

In fact in 2013 wind power alone met a third of Denmark's entire electricity consumption, with that number rising up to 50% into December. On some days, Denmark's wind turbines even produced more electricity than Denmark's consumption required and for one hour on December 1st, the turbines produced 135% of Denmark's demand. In 2014, wind power provided 39% of their total electrical needs.[11]

Denmark got properly going on their low-carbon transition in the 1970s in response to the oil crisis, but their work on wind energy began many years before. As a nation, they were dependent on oil imports as they had limited sources of fossil fuels and limited hydro energy resources. In fact as early as 1891, the government funded research that resulted in the first wind turbine being built. By the 1930s, it is estimated there were 30,000 wind turbines in Denmark. This interest and innovation in renewable energy has continued despite finding oil in the early 1970s and becoming self sufficient in oil by 1997.

A combination of social and political drivers has meant that the Danish people have led the world in the development of wind energy and in many ways the development of renewable energy policy. They were the first to introduce a feed-in tariff (FIT) system that provided premium payments for renewable energy, and they were the first country in 2008 to set themselves a national target to be 100% renewably powered by 2050. This journey has not been without challenges, but has been built on widespread support from an engaged population.

If you travel across Denmark you might want to start the journey playing 'count the wind turbines', but within half an hour you will have lost count, as there are turbines of all sizes just about every way you look. The first turbines were erected by farmers, but soon local communities got involved. It was a case of local people – neighbours to the schemes – buying into a project to provide for their own electricity needs. As these projects grew in size, the pool of investors also increased and widened. In 2002 there was 3,000MW of wind

---

[11] 'Denmark Delivers "One-of-a-kind" Wind Power Record', www.businessgreen.com/bg/news/2388939/denmark-delivers-one-of-a-kind-wind-power-record

energy connected in Denmark, generated from 5,600 turbines and providing 14% of the nation's electricity consumption. Just over 600MW was owned by co-operatives or 'Wind Guilds' as they are known there.

The penetration of wind energy has continued to grow. As we heard earlier, in 2013 a third of the entire electricity needs of the country came from wind turbines, a doubling in the production in just 11 years – revolutionary growth. The percentage of community ownership has dropped over the years but it was the community ownership that provided the stage and foundation for this growth to be possible. People had ownership of the projects so they were happy to see them go ahead.

This strong basis of co-operation meant that projects that would otherwise be very difficult to develop could be built. In 1993, a project was initiated by the Copenhagen Energy and Environment Office who were soon joined by a group of local people. They came together with a view to building a 27 turbine offshore wind park. At the time offshore wind technology was in its infancy, with very few offshore turbines in operation globally. Not put off by the technical complexity of the project, by 1997 a co-operative – the Middelgrunden Wind Turbine Co-operative – had been formed. It raised 30,000 upfront subscriptions of €7 each from 10,000 people to help fund the project development costs. In 1999 a full environmental impact assessment was carried out, along with a third public hearing, after which approval for the project to proceed was achieved with very little opposition. The project was redesigned in a slightly reduced 20 turbine layout to take into account people's concerns, and it commenced construction in 2001. The co-operative owns 10 of the turbines and Copenhagen Energy (the local utility owned by the Municipality of Copenhagen) owns the other 10. There were over 8,500 investors in the project who received a return on their investment of about 7.5%. This is a huge and complex technical project that was developed by individuals and funded by local people, who also make a good return in the process. It is a project where a local co-operative worked with a local authority to create something groundbreaking that works for all involved. It is a pioneering model of community participation and collective action that surely can be replicated in other places.

The Danish wind energy revolution has been at the forefront of creating the global wind industry. Many thousands of jobs and businesses were created in the process, and Danish wind engineering is now harnessed all over the world. Ninety percent of all offshore wind turbines have been built by Danish companies to date. Turnover for the Danish wind industry in 2012 was around €11 billion with about €7 billion as export revenue. It also employed over 28,000 people in the same year.

By 2020, it is expected that 50% of the country's electricity supply will come from wind energy alone, and it seems that they are well on their way.

But community energy and renewable energy in Denmark are not only wind based.

The Danes have also been pioneers in district heating. In Denmark, district heating schemes supply 62% of Danish households – around 1.6 million homes – with heating and cover around 49% of space heating demand in all buildings. Fifty-five of these schemes are run by municipally owned district heating companies and they deliver around 62% of all district heating in the country. The rest are predominantly consumer owned co-operatives with a handful of private companies involved also.

Most of these schemes have been retrofitted to existing buildings and in existing towns during the last 30 years. Many of them are co-operatively owned and run. During their development households were taken off their individual boilers and plugged onto the heat network. The heat pipes are installed around the town and each house is connected via a heat meter, so you get a bill for the amount of heat you use. Each house was charged for this connection, and by law they had to connect and pay the connection charges, but moving forwards they were plugged into a system that they were part owners of. The companies that were established are not for profit, but for the efficient delivery of heat. People connected to these networks pay less for their heating than if they were using gas or oil-fired heating, and the structure of the companies mean that any savings are passed on to the consumers in bill reductions. What is more they are no longer responsible for maintaining a boiler year on year.

These district heating schemes utilise heat and cooling from a wide range of energy sources such as combined heat and power plants/cogeneration (CHP), surplus heat from industry, large solar thermal systems, geothermal heat and large-scale heat pumps. These interconnected systems mean that variable sources of fuel can be used to run the scheme, and that means picking the best priced fuel to achieve the end result. Interestingly many of these district heating schemes are being fitted with large-scale solar thermal plants. At this moment there are some 330,000 square metres of large-scale solar thermal plants in Denmark – creating some 230,000kWth per annum. Solar thermal schemes plugged onto district heat networks provide trouble-free cheap heat, and in many cases they are combined with large-scale storage systems.

THE SAMSØ STORY

The island community on Samsø embarked on a journey to go 100% renewable in 1998. It all started with a response to a competition run by the Ministry of Energy which requested a bid from island communities (there are over 400 in Denmark) to present a plan as to how they would transit to being 100% self sufficient in energy. Samsø drew up a plan, won the bid and then faced the challenge of translating the concept into real action and project delivery in a conservative rural community – whose 4,000 strong population displayed a healthy mistrust of anything from the big city.

Søren Hermansen, an islander from a local community-minded farming family, became the first employee of the project and set to work to try and realise the vision:

I knew how to make the community thrive and also how to make my own position worth something in the local community which is one of the glues in society; that is that you do your utmost and your best at what you are good at in the community for two reasons. One is to make your bread and butter and make a living out of it and the other is to get a position in society that is worth something. I think that is very important that the local community uses or sees the potential of every citizen and that the citizen on the other hand shows and offers his potential for the better of the community.

So, knowing this, and then having the position of being the first staff of the energy island, which was a top-down project from the government, made me convert this from a top-down academic exercise from the government, to be a local bottom up approach project suited to many different purposes in the community. The beauty of this is that energy is a common good that everybody wants and needs and is able to produce and is so easy to talk about because everyone is connected to energy in many different ways. Talking about this makes sense.

People would try to disconnect immediately if I try to say that this is the kind of thing you do out of political, moral reasons. This is something we do because there is a 'what's in it for me?' element to it. That suits many different purposes – the blacksmith, the farmer, the citizen, the hippy – everybody could see their own role in it.

It was not just a 'show off' project, it was more like a 'can do' project that helped people to understand that we were not just customers in an energy shop, but on the other hand we were actually 'prosumers', producers and consumers at the same time. I like the energy of that. I really enjoyed being able to present these ideas to people and also to be part of it. Being in a position where I could call people from the mainland who knew about engineering and could do the planning exercises for the wind turbine and then afterwards call all the local people and say 'here is what we know, so do you want to get on with this and see how far we can go?'

The first thing we did was not actually a project, it was kind of introducing the whole concept of the energy island. We produced a master plan where we indicated how we could be 100% self-supplying. I travelled around in local meeting houses and community town halls and stuff like that and promoted this and talked about it. And told people that this is what we are looking at. I just asked them a simple question, 'What do you think about it, do you think it is something we should go deeper into?' A lot of people said 'that sounds nice, it's interesting', but they couldn't at that time see themselves in it. So we needed a project where we could see ourselves do something.

An energy efficiency programme that accessed a national grant scheme to fund some of the works was where they started:

> So I used this as a door opener. Met with 'Yes, but that's not for us, too much paper work', then I offered and said, 'I can do the paper work for you'. Because people have this feeling that it is complicated to go to the bank. I also prepared the bank and said to them we will probably need... so much per house. When you have done this you, per year, will save a lot of money because you will not use so much energy, and this money may be used for investment in new projects. Maybe we can release some of the investment capital so you can get your hands on shares for the wind turbines. This also helped the carpenters and the plumbers, they like us very much because we gave them jobs.
>
> The grandmothers became my ambassadors, saying, 'you should ... listen to what this guy is saying, this makes a lot of sense, it is good house-keeping to take care of your resources and this is in line with what we have tried to teach you.'

By the year 2000 – in just two years – the islanders had organised themselves enough to have developed, financed and built the first of 11 x 1MW turbines in three separate clusters, which would provide for their entire electricity needs. The rest of the turbines followed shortly afterwards. They were developed following a series of public meetings and with strong local interest in investment, particularly from some of the local landowners. There was already a site on the island that hosted two older turbines and the owners of this project were persuaded to have the site redeveloped with larger more modern turbines in exchange for some of the project. 'Samsø Wind Energy' began pre-registration for the projects and effectively sold two of the turbines to 430 investors in advance. These two turbines remain in co-operative ownership with the remaining nine being held by farmers or groups of farmers.

There were a number of challenges along the way:

> I said to the farmers that's fine you have the land and you have the interest in this but it... if you run away with the whole ownership... people will probably hate you for the rest of your life because they will be sitting out there on the porch looking at these ugly wind turbines, spinning money into your bank account and not to theirs so why don't you just come up with a suggestion so they can own part of the project. Of course we had some problems along the way because there were some areas that people didn't want them, they... didn't like the sight of them. It's perfectly OK to

be negative… and say your opinion straight in a very frank way. Where do they harm the least? How can we agree on that?' That was a very healthy discussion for the community.

So that was the first of the three of the 1MW turbines that was up and running in 1999. We sold all the wind turbines upfront before we actually bought them on a pre-signed contract so we knew they were fully financed. And then the next two sites with another three and another five, so altogether 11MW was built in 2000 and 2001 so that was very quick.

To effectively offset the carbon emissions caused by transport on, and in travelling to and from, the island, the Samsø Commercial Council, Samsø Farmers' Association, Samsø Municipality and Samsø Energy and Environment Office joined forces and founded the Samsø Offshore Wind Company, to lead a complex development to build a number of offshore wind turbines. This way they ensured participation from a wider range of potential partners and were also able to draw on their strengths.

That was a huge project and we were very nervous whether we could handle that. So we asked around, all the Shell Oil Company (Dutch Shell) because they had a green policy also about Green Development. They said they would like to be part of that project if they can own everything, all the wind turbines. So we said 'No, no, no, you can maybe have one or two', but they didn't want to be part owner.

It was kind of a struggle because we needed to ask the State. The sea bed is owned by the Crown or the State so we needed to put in a commission quite early in the process just because it takes a lot of time to get planning permission for this.

In 2002, work started on building 10 x 2.3MW offshore turbines, which because of their placement, generate substantially more energy per MW than the onshore ones (3500MWhr vs 2300MWhrs per MW installed). Of these turbines two are owned co-operatively by small shareholders, five are owned by the municipality, and three are owned privately.

In areas where there were collections of dwellings, the energy masterplan and projected district heating systems would be used to provide heating. In fact there was already a district heating scheme on the island in the Tranebjerg community serving 400 consumers. Established in 1993 and run and owned by the local utility it runs on straw, chopping the bales before blowing them into the boiler – so the boiler can run at lower outputs.

Following various public meetings where the idea was explained and sold to residents, a second district heating system was turned on in 2002. Serving 178 heat consumers including some industrial ones, this system was interesting because it used woodchip and solar thermal panels as the primary fuel.

In order to get a project like this underway the group of people developing it had to first establish that enough of the households in the area wanted to make it happen.

The project was planned to serve heat to two villages. In 1999, more than 70% of the people signed up and paid a small fee (100 Danish crowns) to join the project at this early stage. If they waited to sign up they would be liable to pay the full cost of connection (40,000 Danish crowns) – so this gave people a real driver to get on with it. The woodchip for the plant is delivered from the south of the island, and the 2,500 square metres of solar thermal panels provide 25% of the annual heating requirement of the system. The plant and infrastructure is commercially owned by the local utility, who had experience of running systems like this already.

The first district heating started to be planned in 1999 and in 2001 it was sold and planned and I think in 2002 the first plant started. It took about one year to do the planning. We needed about 70% of all house owners in the district to sign the contract. In the beginning we didn't get it at all. We had 30% or 40% for the first one. We were pretty frustrated. People thought it was a good idea but didn't really trust district heating. They were pretty sure it was going to be very expensive.

We made a few mistakes. One of them was to believe that they would trust what we said and they would probably buy this idea, which they only did theoretically. We couldn't make them sign a contract.

We tried to break it down to how many man hours and much equipment was put into this. Then we started talking to the plumbers, the installers and the builders and the contractors and evaluated and estimated how many man hours of work there was embedded in the project and said to them, 'This is your bread and butter, this is your business, we are not here to tell you that this is the way to do it but you could be in your position to make a project like this.' And then they started talking in a positive way about this and they became ambassadors. Gradually they won over the citizens one by one because if you don't trust the blacksmith who do you trust? And that was quite interesting that it helped so much that it was the local blacksmith that promoted the idea instead of even me being a local – they still believed me to be a little bit of a hippy and a little bit too optimistic, but when the blacksmith said, 'This is what we'll do and it's going to be good', everybody knew this was a guy who was talking about his own business but also what a guy who could mend and fix it if it didn't work. Then we also went to the farmers and talked to them about supply and feed stock for the boilers. We had some discussions about the price, and we made a five year contract with them. So then we could go out to the customers, 'We have a five year contract, we know that energy price per MWhr will be so much, oil price today looks like this and we predict that the price will go up a little bit every year for the next few years.' Even with this math they could save up to more than 20% of their present heat device. That was a very convincing argument.

There are now two more district heating schemes on the island also running on straw so that 700 heat consumers in total are served by the district heating networks. For the remaining dwellings that were off these heat networks, work was undertaken to look at people's homes and help them access funding to improve the efficiency and install renewable energy systems like solar thermal and biomass systems. This work meant that the project grew trust with the wider community.

Amazing successes achieved in just 10 years. Of the success Søren says:

It feeds good and very fulfilling. I can see it worked and still works, but in the process I think there is much more to say. It has been full of frustrations, periods where people were negative and even a bit aggressive because of me and my colleagues pushing maybe too hard. Us being very much aware of national policy and possibilities, them being worried about their next job, income, business etc.

When things worked it was mainly because we were in line and able to say WE and not them and us. I was thrilled to take the floor in the Japanese Parliament and tell them the story about a community that decided to be the change. It is great to participate in a local meeting and feel the communication changed to a more holistic and inclusive one instead of the old conservative protective version we came from. It is still there but we know it now and we do not feel opposition and criticism as much as we did before.

The power of this transformation is spreading through all levels of Danish society. Now they have a serious target as a nation to be the first fossil fuel free nation. By 2020, if all goes to plan half of Denmark's electricity will be produced from wind and by 2030 coal will be phased out entirely. It is hoped by 2035 all of Denmark's energy demands in electricity and heating will be met by renewable energy. To complete the journey it is hoped by 2050 all of Denmark's energy will be clean, safe and renewable. They are well on their way.

CHAPTER 9

GERMANY

In Germany, at midday on Sunday 11th May 2014, the combined output from wind, solar PV and other renewable power sources peaked at 73% of the total consumption production. And, during the first quarter of 2014 wind, PV and other renewable energy sources delivered 27% of the total electricity demand. On that day power prices on the wholesale market went negative for six hours.[12] This low demand from conventional power plants had an impact on the electricity prices not just in Germany, but across Austria, France, the Netherlands and Switzerland as well. In 2012 about 23% of the entire electricity use in Germany came from renewable sources – a number which had actually doubled in six years. In 2014 that figure had risen again to 25.8% of all of their electrical needs coming from renewable sources, so the transformation is continuing.[13]

The movement to transform Germany's energy system started with the oil crisis in the 1970s, and whilst these figures from 2014 may not seem like a massive achievement, they come from the fourth largest economy in the world, one that is famous for its industrial productivity. So with that in mind they are a great example of what is achievable in an advanced industrialised nation.

Germany has been heavily dependent on imported fossil fuels, as lignite coal is one of the few local resources. As three quarters of the fossil fuel energy is imported, the roll out of renewable energy has the potential to massively reduce the spending on imported fuel. In 2009, electricity, heat and fuels from renewable energy saved imports of fossil fuel based energy sources worth €5.1 billion – a massive saving.

The path to renewable energy reaching this significant role in the energy mix was catalysed by the Chernobyl disaster in 1986 and the increasing awareness of climate change that was just emerging at the time. People responded to Chernobyl by looking to phase out nuclear, and organising themselves to work out how to achieve that on a local level as well as politically.

In 2000, renewable energy in Germany started the journey of going mainstream with the introduction of the Renewable Sources Act. This act brought

---

[12] *The Energiewende in the Power Sector: State of Affairs 2014.* 2015, Agora Energiewende.
[13] 'German Power Prices Negative Over Weekend', www.energytransition.de/2014/05/german-power-prices-negative-over-weekend/

in the now world famous feed-in-tariff laws that give a higher price for energy from renewable sources, thus making them profitable. It set out provisions to simplify the process between generators and utilities for installing renewables and built on earlier law that gave priority access for renewable energy to feed into the grid. By 2011 there was 65GW of wind and solar PV on the grid – the highest deployment of solar PV by any nation on earth. In fact the 36GW of solar on the German grid in 2012 actually represented about one third of the entire capacity installed globally.

The 'energiewende' or 'energy transition' as it has become known has become a powerful force and a national icon. The momentum to deploy this level of renewable energy has involved many sectors of society: individuals, businesses, villages, municipalities, cities and federal government. There are so many pioneering communities, towns and projects it is hard to know where to start in telling the story.

SCHÖNAU

The 'Stromrebellen' or 'power rebels' of Schönau are a good place to start as this small Bavarian town has been a source of inspiration to the nation as it trailed a pioneering path in the energy revolution. The 'citizens' movement' in this town of 2,500 people started with a desire to be free of nuclear energy post the shocking events of the Chernobyl disaster. What started with energy efficiency campaigns and requests to the local utility to stop sourcing energy from nuclear power stations, took a leap forward in 1997 when the grid system in the town was purchased by the community and brought under local control. By 1999 the company that had formed to achieve this, Elektrizitätswerke Schönau (EWS),[14] thanks to deregulation of the markets was able to sell electricity to people all across Germany.

I met with Tanya and Eva from EWS to hear about the journey they had been on over the years:

In Southern Germany, the irradiated rain clouds came, and people were recommended not to eat salad or to drink fresh milk. It stopped after a few weeks but Michael Sladek (one of the founders of EWS) is a doctor, and so he was very aware of the problem and it's not a problem that lasts for just a few weeks. Because of the radiation people got in touch with each other to get an idea of how to avoid irradiated food. They came to the playground and took out the old sand, the irradiated sand.

They started sharing information and they collected donations for the hospital in Kiev to buy medical supplies. The parents of Schönau invited kids from the region of Chernobyl to the Black Forest, to assist their

---

14   www.ews-schoenau.de

recovery. They did it for several years in a row and the same group came, year after year. One year the group had one less child because one little girl had died from cancer. We realized that this was happening not far from us and their children looked the same as our children do.

The group which formed had been named Parents Against Nuclear but we realized we had to find alternatives, we had to be in favour of something – so we became People for a Nuclear Free Future.

In response the community that formed began running an energy saving competition, with prizes for energy saving achievements.

When we started energy saving competitions and awarded people for saving energy we didn't make friends with the local energy supplier. We approached the local supplier and asked them to hire us meters to help improve our energy saving attempts, but they were very arrogant and would not help. The supplier was powerful but saw his business being destroyed and this was the time when we realized, 'We have to become an energy supplier ourselves'. We cannot co-operate with the monopoly because it is not their interest to save energy and they don't want to change. So we have to do it ourselves.

The supplier approached the town council and it was four years before the license ran out for the grid. The supplier offered 100,000 marks to get a new license. The people had to run a referendum to prevent the town council from giving the license to the supplier. It was a revolution in Germany, the first time that people demanded a referendum for not giving the grid license away. In the very beginning they had the idea, 'If they don't help us then we will do it on our own, but we don't have a clue how', so they just took one step at a time – many things happened by chance. There was a lawyer in the group who knew that there was a possibility to make a citizens' referendum and the council was completely astonished that these means existed. I think there could be potential for this in other places…

But it didn't end there, once the people won the referendum:

The opposite side said, 'No, we want a referendum'. They had to run the second grid referendum and we went on to win the second referendum as well. The people got the license and then they had to work out how to actually buy the grid. They had a quarrel about the price of the grid because no one in Germany could say how much the grid was worth. The old supplier said it was worth 8.7 million Deutsche marks and in the end after seven years of quarrel we paid half the amount.

We had to run a big donation campaign because we were not allowed to give more shares and overestimate the business. So the business was funded by social bank shares, by citizens and then donations. The donation campaign collected two million Deutsche marks. At that point of time we knew that all over Germany the spotlight was on Schönau because it would be an example for others to follow.

People from all over the country donated to the cause – despite the warnings from some not to. One utility from 500km away wrote to all of its employees saying: 'We don't support those guys in Schönau and if we find out that any of you give any kind of donation to their initiative then you will no longer have your job.' Fortunately some people ignored the threats and donated anyway, and the story became a national David and Goliath battle.

Next they actually had to work out how to do it, so they invited experts from all over Germany and even Europe to learn how to become grid operators and energy suppliers. This whole process took quite a number of years with the first referendum taking place in the early 1990s but not commencing operation of the grid until 1997. They only became a supply company over the following couple of years.

> We were a very small supplier as we started with a few thousand customers and we were thinking about how to make our assets as effective as possible. The result was not to build our own power plant, but distribute a small amount of money to help people to build up their own. In this program there were about 2,200 small power plants built mainly PV and co-generation and a few wind turbines. And afterwards when we had more resources we launched the development company which is ESW Energy GmbH Limited and which bought shares in PV parks and other renewable projects.

Today, just 15 years on, EWS has become a significant supplier of renewable energy with the company now serving 150,000 customers across the country. Ninety-five percent of the electricity it provides comes from solar, wind and hydro resources, with the remaining 5% coming from CHP plants. The co-operative has a number of trading companies running different areas of the business and operates seven electricity grids and two gas grids in different communities.

For others setting off on the journey in their community, Tanya said: "Everything is possible, step by step. Be as naïve as possible, that will help you. If you know the kind of obstacles and hurdles you are facing then maybe you would go back to bed, but if you don't have a clue just run forward and the way will open up. And become angry. Anger is a good motivation. Yes, hold the anger!"

FELDHEIM[15]

Another iconic town is that of Feldheim, with a population of only 130 people located about 40 miles south of Berlin, who now use only 1% of the electricity they generate to cover their local electricity needs. In 1995, a local entrepreneur built the first wind turbine in the town, inspiring farmers whose incomes were being squeezed to get involved. A local renewable energy company, Energiequelle GmbH, installed 43 wind turbines across the Feldheim landscape, providing income to farmers who leased their land to the energy company. In 2008, they went on to build a solar farm on a nearby abandoned Soviet military base. Most people you'd have thought would stop there, but this community had got the bug. In the same year, the town established a joint venture with Energiequelle GmbH which went on to build two biogas plants that convert pig manure, cow slurry, maize and wheat into biogas. This biogas is burnt in a plant that produces electricity as well as heat. Reusing the waste from the community's pigs and cows, as well as some agricultural products, produces 4 million kWh of electricity a year, $11,500m^3$ of fertiliser from the digesters which goes back to the agricultural co-op closing the circle, as well as heat to the community buildings. These plants also supply the heat via a district heating network which has replaced the use of 160,000 litres of heating oil every year. This is supplemented by a 400kW woodchip furnace fuelled by the by-products of forest thinning.

By 2009, Feldheim was producing all of its own energy with renewable sources. As a next step, residents then attempted to buy the grid from operator E.ON, who were frankly less than keen. Unperturbed, the people decided to build their own new grid to replace the existing one, with each of the residents contributing €3,000 for their new smart grid. Completed in 2010, with support from the EU and Energiequelle GmbH, this allows the locally produced electricity to be fed straight to consumers and gives them control over their electricity prices, which are set at community meetings.

Werner Frohwitter works for Energiequelle GmbH and describes the unique circumstances that led to this commercial public partnership:

> My company owns a large wind farm that is about 1,000 yards away from the village and we have now 43 wind turbines. You do not need 43 wind turbines for a community of 130 people. In fact one turbine would be sufficient. Just to give you an idea, every year Feldheim consumes 1 million kWhrs of electricity and the wind farm produces 140 million kWhrs.
>
> Right from the beginning, the people of Feldheim have been involved in the planning and the management of the wind farm, be it the land lease with the landowners or job creation and maintenance work. The project

---

[15] www.neue-energien-forum-feldheim.de

has become a joint venture with the farming co-op who are an important force in the village. We bought a biogas station about six years ago that gives the farmers more possibility to market their products, not only on subsistence crops but also on energy crops, stabilizing their income.

The relationship between my boss, Michael, and the inhabitants of Feldheim grew steadily and it has now developed into a very deep trust for one another. He always had it in mind to build a factory, for photovoltaic parts, and of course it was obvious that this factory was placed in Feldheim. The idea was to connect this factory directly to the wind farm to get green power not from the official grid.

Then the boss of the farming co-op came up with the idea of connecting his pig farms as well to the wind farm and to the biogas station. After all, the farming co-op grows piglets, 10,000 per year and piglets need a lot of heat in the first and last weeks of their life. So the original plan was to connect the pig farm and the factory to the biogas station in order to get their heat from the biogas station which is a secondary output from the electricity generation as well as the electricity from the wind farm. But then someone from the population came up with a question: 'Why can't you extend the heat pipe lines and the power cables to my house? I am a few yards away from the pig farm.' Of course, a second neighbour became aware of this plan and wanted to be connected as well as a third. Then very quickly a meeting was organized for the community and after a few meetings everybody was agreed that the village was to be connected to the wind farm and the biogas station. The citizens of Feldheim formed their own power utility company and with our help. We joined in and did all the finance and funding stuff, the red tape and the negotiations with the old grid operator.

We had a cable from the wind farm to the village from one string of the wind farm to a new substation inside of the village, but of course this needed to be connected to the local distribution grid and the owner would not sell it to us or lease it to us.

We had two options; one to drop the idea and just to shrink it down to the heat pipe connections, or the other idea was a very intrepid one, to build a new distribution grid; a completely new, smart grid inside the village. Well, we just did this. We dug up the streets and constructed a heat pipeline, a district heating net and a new distribution net. It all took about a year and a half and in Autumn 2010 the system went into operation. It worked smoothly so now Feldheim has electricity directly from the wind farm and heat from the biogas station. No more coal, no more timber or wood or heating oil, just clean energy from wind and biomass.

It was not that easy because it was a completely new way, it was new to the authorities and of course we had a serious problem with the old power company. Just imagine, we break away from the old power company and they weren't very pleased about that because it was a precedent case in Germany. We succeeded with it but since then the laws have changed here so Feldheim is not repeatable.

The electricity for the villagers is about 30-40% cheaper than the average price in Germany. That makes quite a difference and of course local value was created, jobs were involved and income tax and industrial tax were paid. The treasurer of the community is very pleased because every year many thousands of euros come into his accounts.

We will now be installing a battery storage system in the village of 10MW. This is a new idea. We have already secured the funding and work is going on. The system will be ready in about eight months. So then we can store the surplus wind power overnight and sell it during the day at better prices. It is all about money, isn't it? But it shows you that ecology and economy can be reconciled. They are not opposite things. You cannot have economy *or* ecology, you can have both if you do them a clever way and if you stick together, if you have community spirit.

My company were not just the engineers, we were executers of the plan, but the plan was born in the community, the citizens brought forward the idea. They gave birth to the idea, they joined in with their money and with our help it was possible.

Feldheim Community Energy Company owns the grid and all the paraphernalia of the grid, the substations, the house connections. They purchase power from the wind farm owned by my company so we sell each kilowatt hour the same price we would get if we exported it to the national grid. Feldheim is purchasing their energy for eight cents and sells it to its customers (which are the community themselves, every partner is a customer, every customer is a partner), at the price of 16.6 cents. This makes eight cents to fund the grid construction and finance costs and to make a very small profit because you have to pay taxes. What is not included in this price is a big profit. Once the grid cost is paid off in 12-15 years, there is a margin to reduce the price to an even lower level.

The bigger the community, the more difficult it gets because every single household has to agree. You need a public-private partnership because if you are in a normal community how do you run your own power company? How do you secure the power supply when there is no wind? You need specialists. You need electrical engineers, you need funding specialists, you need bankers, so you need to join quite a lot of forces together and there is hardly a village who has all these talents within its population. What you need is a private-public partnership like we did with my company and the community of Feldheim. You need to get organized and getting organized is a complex problem.

We have a saying in Germany, 'Your own pig doesn't smell.' The neighbour's pigs smell but my pig doesn't. So make it their pig. People need to be involved not only in the planning but also economically. You will always have detractors of course but if you convince people of the ideas and let them participate economically as well, they will say: 'This is my pig, my power and I'm earning with it.'

Feldheim is the only town in Germany with its own micro-grid set-up, and the whole process has not only secured lower and stable energy bills for the community, but it has also created local employment. It has helped to bring prosperity to a rural community when others around are struggling. This little town has become a popular destination – with no museums or attractions, people are coming from all over the world to see the energy revolution in action.

Community ownership of the renewable energy assets that have been developed has been a common theme in the roll out of renewable energy across Germany, and the numbers of co-ops have been growing at an amazingly fast rate. In 2006, there were only 86 of them, but by 2013 this number had risen to 888.[16] These growth rates mean that a new co-op springs into life in Germany about every three days. People are investing in these schemes for a number of reasons: to invest in something local and tangible with a steady rate of return, but also to be part of creating the energy transition.

Germany's energy revolution has not been isolated to rural communities; people in towns and cities have been heavily involved in the process. In fact they have not only got engaged with building and owning the generation equipment itself, many communities including the country's largest cities have attempted to buy back the network – the power lines, transformers, substations and district heating networks – in an attempt to speed up the process.

In Hamburg, Germany's second largest city, a referendum was held in 2013 that voted by a 51% majority to buy back the grid from the utility operators Vattenfall and E.ON.[17] The diverse consortium that formed to champion this cause was made up of more than 50 environmental, consumer and church groups. Their reason to argue for this change was that the private utilities don't act in the best interest of the people and are delaying Germany's shift to renewable energy.

The mayor of Hamburg actually opposed the move, but success in the referendum means that the municipality now has to form a locally owned utility to take control of the network. Germany had a similar policy of privatisation of the electricity system to many other areas of the world during the 1990s, and it resulted in an increase in electricity prices from 1998 to now.

The pro campaign in Hamburg estimates that the municipality running

---

[16] 'Community Ownership Energy Co-operatives Up, Up, and Away', www.renewablesinternational. net/energy-cooperatives-up-up-and-away/150/537/76751/

[17] www.bloomberg.com/news/2013-09-23/hamburg-backs-eu2-billion-buyback-of-power-grids-in-plebiscite.html

the local grids would generate €1 billion in sales and a profit of €100 million a year in grid fees. They will take on 27,000 kilometres of power grids, 7,300 kilometres of gas pipes and 800 kilometres of district heating network. The campaign centred on the concepts that leaving these networks being run solely on a for profit basis would not enable them to be transformed to cope with the roll out of distributed energy, and that energy should not be a commodity sold simply for profit but a force for common good as well. However the grid in Hamburg, as in many other places, is old and it will need significant investment to ensure it remains strong moving forwards. This will inevitably increase the prices to consumer and may prove a challenge.

Berlin went through a similar referendum in 2013 in an attempt to buy back its grid, which failed by a narrow margin. However since 2007 about 170 municipalities have bought the grid in their areas from private companies.

The 'energywiende' in the electricity sector has been paid to a large extent by a levee on consumer bills, and the cost of it is actually displayed on the bills themselves so individuals can see how much it is adding to their costs. This has reached quite a high level in recent years, in some cases as much as 20% of the bill. This is partly due to the fact that there are a range of businesses, the heavy energy users, like manufacturers, who are exempt from the levy. So the burden of it is loaded onto consumers. Germany has one of the highest electricity prices in Europe, but people's bills are significantly less than in many other countries as energy is generally used more efficiently. One might think that this increase in bills would make the process unpopular, when quite the opposite is the case. The 'energiewende' was actually brought about by public protest and surveys of public opinion over the last two years show that 90% of the population felt that efforts were 'important' or 'very important'.[18]

Whilst this may seem an altruistic act on the part of the German population, it is also a shrewd move in terms of future proofing the country from the effects of fossil fuel price increases. The economic costs of the environmental and health related risks of nuclear and fossil energy are reduced. Whilst at the same time embracing the potential of energy efficiency, renewable energies have created many local jobs and export opportunities. The heavy investment made in solar PV particularly has driven the massive cost reductions seen in the technology during the last few years that the rest of the world will benefit from. For Germany, investment now will reap rewards for many years to come in the German economy. Already in 2013 the sector was worth €40 billion and employed over 400,000 people.

With these numbers of people so actively engaged it is not surprising it has such widespread support. However it didn't start that way and the massive growth in renewable energy has been a hard-fought battle, won in small steps by policy makers, NGOs and communities all around the country. In the

[18] Piria, Raffaele (2014). *Greening the Heartlands of Coal in Europe.* Heinrich Böll Stiftung. p.21. www. boell.de/en/2014/03/18/greening-heartlands-coal-europe

process, the political environment has been changed to facilitate further development of the sector. Part of this change saw the announcement of the planned closure of all nuclear power stations in Germany by 2022 – what those 'Parents Against Nuclear' had been seeking all along.

During the period 1991-2011, Germany's carbon emissions dropped by 24%, whilst their GDP grew by 27%.[19] Despite the huge success of renewable energy and community power the struggle is far from over. Collectively, German citizens have shown what is possible in such a short period, and are still taking to the streets to demand more change faster.

The challenges moving forwards will be how they can replace their use of coal with renewably generated electricity. As the penetration of renewable electricity on the grid increases, solving the issues around storing electricity will become even more important, to avoid shutting down solar when the sun is shining. It seems they are rising to the challenge as a nation despite all the naysayers along the way.

One thing seems certain though, the pioneering approach will continue. As the tariffs that have supported this renewable energy growth are being brought down, business models and approaches will inevitably have to change. One pioneering co-operative in Thuringia, called EG Rittersdorf,[20] has built a 1.5MW solar plant and teamed up with a Hamburg based utility to start providing power directly to local people. The idea being that the utility pays the co-op slightly more than the feed-in tariff rate for the power from the array, and in turn markets a '25% solar tariff' available to local people. This tariff is competitive and only available within 30km of the installation. Their calculations show that once they have 1,500 customers signed up to the plant then they will no longer be dependent on the feed-in tariff. As a first step on the transition to subsidy free, this seems like an elegant way to find a financial model that works whilst bringing people along.[21]

On the other side of the story, towards the end of 2013 RWE announced that it was taking 6% of electricity generation offline, saying: "Due to the continuing boom in solar energy, many power stations throughout the sector and across Europe are no longer profitable to operate."[22]

This surely demonstrates the powerful disruptive force that the roll out of renewable energy, in broad ownership, can have on the existing energy infrastructure of our developed nations. The 'energiewende' has become the fertile ground for new ideas to flourish, disrupting the incumbents, and putting power back into the hands of the people.

---

[19] 'Hendricks Strives for More Ambitious Climate Targets', www.umweltbundesamt.de/en/press/pressinformation/hendricks-strives-for-more-ambitious-climate

[20] EG Rittersdorf, www.engeri.de

[21] 'German Energy Co-operative Moves to a Direct Market Model', www.solarnovus.com/german-energy-cooperative-moves-to-a-direct-market-model_N7608.html

[22] www.bbc.co.uk/news/business-23692530

CHAPTER 10

UNITED STATES

By May 2014, there was enough solar energy deployed in the US to power three million homes. During the first quarter of 2014 solar accounted for 74% of all new installed electrical generating capacity. A big chunk of the installations that make up this figure were comprised of domestic home solar systems, of which one third were built without any state subsidy.[23] The speed of roll out of solar in particular can be quite astonishing. In 2013, the state of California installed as much solar capacity across its homes and businesses as it had done in the previous 30 years combined, going from 1GW to 2GW of installed capacity.[24] Around 200,000 US homes and businesses added solar in 2012 and 2013.[25]

Meanwhile in Texas, a state well known for another sort of energy, there is a wind power boom. In fact on Wednesday 27th March 2014, a new record was set with wind power providing 29% of the state's electricity needs. Surprisingly Texas has the most wind power of any state in the US, with wind providing 9.9% of the total energy needs in 2013.[26] Even T. Boone Pickens, a well-known Texas oilman, corporate raider and takeover specialist, is on board with wind in the state.[27]

Great renewable energy headlines aside, the US is still one of the world's most energy-intensive countries. It was the world's largest consumer of oil in 2012, consuming nearly twice the volume used by the people of China. Total energy consumption in the US is second only to the Chinese even though per capita consumption is over three times higher in the US. Oil and gas production has dominated the history of the country in the last century and right now renewable energy only accounts for roughly 14% of electricity supply and less than that of total energy consumption.

---

[23] 'Solar Market Insight Report 2014 Q1', www.seia.org/research-resources/solar-market-insight-report-2014-q1

[24] 'More Rooftop Solar was Added in California in 2013 Than in the Past 30 Years Combined', www.treehugger.com/renewable-energy/more-rooftop-solar-was-installed-california-2013-past-30-years-combined.html

[25] 'Utilities Feeling Rooftop Solar Heat Start Fighting Back', www.bloomberg.com/news/2013-12-26/utilities-feeling-rooftop-solar-heat-start-fighting-back.html

[26] 'One-Third of Texas Was Running on Wind Power This Week', http://motherboard.vice.com/read/one-third-of-texas-was-running-on-wind-power

[27] www.pickensplan.com/the-plan/wind/

However the US has a long history of energy activism, innovation and engagement. In fact probably one of the most forward thinking of those energy activists was President Franklin Roosevelt. As part of his 'New Deal' in the 1930s, he signed a piece of legislation called the Rural Electrification Act of 1936. At that time 90% of people in the cities had access to electricity, however 90% of the rural population did not. (This reminds me of the grid infrastructure challenges found across Africa today.) The investor-owned utilities that served the cities couldn't make the returns they were seeking by building distribution and transmission grids for rural areas so they simply didn't do it. Those in rural areas who did have a supply were often paying much higher electricity rates than those in the cities. In 1935, and in the face of opposition, Roosevelt created the Rural Electrification Administration. This body went on to design the rural co-operative utility model that was then rolled out. It provided a low-cost and long-term loan programme to finance the required development and also provided technical, managerial, and educational assistance. By 1939 there were 417 of these rural utilities serving 288,000 households.[28] An amazing achievement in such a short space of time.

This intervention paved the way for the current electricity system in the US. Electricity is provided to consumers by a range of different businesses: investor-owned utilities (IOUs), municipal utility districts (MUDs), public utility districts (PUDs), and co-operatives. There are around 3,200 companies active in the sector, and many are actually owned by the people they serve. There are still 900 electrical co-operatives which are owned by their customers and 2,100 cities or counties that own their own utility company. Generally, these collectively owned utilities have only dealt with the distribution and transmission side of the business while 75% of the power generation is provided by the IOUs. On average, public power costs about 10% less than electricity provided by private companies.

So today quite a number of Americans are provided electricity by non-profits that have a management board electable by its members. There are some interesting and innovative approaches to the roll out of renewable energy technologies across the spectrum of companies operating in the sector from both the IOUs as well as the other providers.

PIONEERS OF RENEWABLE ENERGY

I find it amazing that in 1888 in Cleveland, Ohio, an engineer by the name of Charles F. Brush built what is believed to be the first automatically operating wind turbine for electricity generation. The turbine was a giant with a rotor diameter of 17m and 144 rotor blades made of cedar wood. Brush was an energy pioneer with a string of patents to his name whose home power system

28   Electric Co-ops History, www.se-coop.com/coops_history.html

was the first electrical system in the area. Producing 12kW of power this monster system ran his home for over 20 years with batteries in the basement providing a consistent supply.[29] What captured people's imagination was the fully automated nature of the system – something that hadn't been seen before.

Fast forward to the 1970s and there were still people out there who had no access to the grid and electricity as they chose to live in remote locations. There was another pioneer who set up shop to support those people to access energy from renewable sources. John Schaffer, founder of 'Real Goods', became the first person in the US to sell solar panels to people for powering their remote lifestyles. In many ways he is the grandfather of an industry. For me Real Goods' *Solar Living Sourcebook* – now in its 30th year edition – is a bible that I spent many hours poring over and then used to actually build my first solar system. It is still as relevant and useful today as it was when I first picked it up 20 years ago – if not more so.

John has had the unique experience of seeing solar, and renewable energy in general, evolve from nothing to where it is today. I asked how it felt to be the first person to sell a PV panel in the US and possibly the world:

> Well I must say it feels great. It is a little conflicting from a competitive business standpoint. In the beginning, from 1978 until about 2003, there really wasn't any solar business because there wasn't any net metering or any incentives anywhere, so we had a monopoly on it but there wasn't a business big enough to make it go. So now when you see, going from 5GWs to 10GWs, expanding outrageously it is very heartening to know we played a part in getting all that started way back. It is easy to get cynical and think we paved the way through the jungle with machetes so that all these other companies could drive through in their SUVs to get to the same place when we had to put in all the sweat. But in balance I have the satisfaction of knowing that we helped start something that needed to be started.
>
> In the 90s when we started a billion pound carbon dioxide goal, where we declared in our catalogues that we were going to eliminate a billion pounds from 1990-2000 and we began to measure every compact fluorescent, every photovoltaic, every water saving shower head we sold. Then we began to publish in our catalogue how that converted into kilowatts saved and prevented CO_2 from being emitted. That was a really satisfying thing to see; how that opened the eyes of the public when they had never heard of global warming.
>
> Back in the day there was no solar market really, it was all people in the back woods at the time that didn't have any access to power. So right around 1997 or '98, when Pacific Gas and Electric (PG&E), our local utility, decided that net metering made sense, my wife and I put up the first

---

[29]   'A Wind Energy Pioneer: Charles F. Brush', www.motiva.fi/myllarin_tuulivoima/windpower%20web/en/pictures/brush.htm

grid tied PV system; an entire array in Northern California. I was getting kind of depressed that this market was never going to really take off – we can sell wood stoves and farmstead equipment and books but the solar technology just wasn't getting going. Then once PG&E made that jump into it and you saw one or two states and then 10 states and now there is 42 states that have net metering. That was a real shot in the arm, knowing that success could come around.

That was one of the first things that helped it go mainstream and then the second big one was the advent of the power purchase agreement and leasing agreements, where instead of people buying a system themselves, they got a system on finance. Of course the price had fallen, just three or four years ago we were selling systems for $8 a watt. For the average home 5kW system, they had to come up with $40,000 in cash and then various banks and credit unions came up with a way to get down to zero and someone else would own it. Over 20 years, you would pay less money than you would pay your electric company. That's clearly the biggest thing that has happened in the last two or three years that has taken it mainstream. It is no longer an environmental decision, it's an economic decision. The prices have come down just in three years over 50%. As I say, it was $8 dollars a watt and now they are down to $3 dollars a watt.

John and his team have helped thousands of communities and individuals across the US and further afield to use renewable energy. What motivates people to get involved?

Usually it suffices for them to see it is economically feasible. It's not a pipe dream, it can actually save them money so it is not a very hard sell any more. There is this new movement, called Solar Gardens, where people who can't put solar directly on their home will have a community-owned system that they'll each contribute to and they get some kind of a credit for doing it. There's a big one in Colorado and now they are moving in Massachusetts, so it's a way for people who live in apartment buildings to have solar too.

Change seems to be accelerating every year, almost doubling. It seems like in another 20 years or so we could be at 40% or 50% penetration once the price continues to go down and people realize that fracking technologies and coal technologies are all disastrous. It is only going to go up and up and up so I am pretty optimistic about the future.

Optimism is infectious, and breeds creativity. It seems there is an explosion of creativity around energy and particularly community energy going on across the US today. The range of approaches taken by communities, municipalities and businesses adopting renewable energy is as diverse as the country itself.

There are however some key themes and models that are repeated across the country.

THE PEOPLE BUILD THEIR OWN RENEWABLE POWER STATIONS

It seems that the co-operative model for development of community energy systems that is widespread in Europe is not so common in the US because of various financial and legislative barriers. However it is being done. Schemes are also being developed with broadly the same principles by dedicated groups of volunteers. Instead of a co-operative structure, people are using Special Purpose Entities (SPE) or non-profit models to run their projects. Doing the same work as many European based co-ops, the SPE companies are set up for people to invest in solar or other renewable energy in their area.

Sometimes they are set up by brokers, but often by enthusiastic individuals. One example is Greenbelt Community Solar, LLC set up by a group of residents to establish a solar power generation plant on the Greenbelt Baptist Church. They began generating electricity from a 21.6kW solar photovoltaic array mounted on the church roof with a 20 year power purchase agreement to ensure returns for investors.[30] There are many others across the country. However they are limited in that only a small number of investors can be involved in each SPE so the ability for large numbers of people to participate is reduced.

'Community solar gardens', however, seem to have found a way to give people access to solar despite these issues. They are set up as a way for people who otherwise could not access the technology, because they live in a flat or a rented home, to do so. The model is to set up a co-operative as well as an SPE. The subscribers may purchase a portion of the power produced by the array and receive a credit on their electricity bill. This is either achieved by buying an actual stake in the project, making a return on that stake over the course of the investment, or by joining the scheme with no upfront investment, and getting a reduction on their monthly electricity bill. Specific 'solar garden laws' have to be passed to enable their creation and so far this has been achieved in Colorado, Delaware, Maine, Massachusetts, Vermont, and Washington.

The Brewster Community Solar Garden® Project in Massachusetts constructed a large solar array in a field adjacent to the Brewster Water Department. The site, in Brewster's industrial zone, is a former sand pit and disposal location which would be unsuitable for most other development, so perfect to host the array. As Kit Reynolds, one of the founders of the project, puts it: "We are in an unprecedented period when it comes to energy choice. We can choose to invest in solar, keep our wealth local, provide jobs for local workers, move toward energy independence and improve the environment.

---

[30] Greenbelt Community Power LLC, www.greenbeltcommunitysolar.com/home

We can do these all at once."[31] Win-win.

There are yet more opportunities for win-wins. Getting involved in generating your own energy is not limited to richer communities that might be seeking opportunities to invest, there are programmes that reach out to low-income households. Grid Alternatives is one of the pioneering organisations making sure of that. They do a mixture of win-win things; they provide low or no cost solar to local homes, whilst at the same time offering volunteering as well as training programmes for people in the area to get hands on experience of solar. As the founders' state on their website: "Together we have built a rooftop power plant that is putting millions of dollars back into the hands of families with the greatest need." In their first 10 years, they've built 14MW of rooftop solar spread over many homes, and they plan to expand that to 100MW over the next 10 years. They have funded this impressive work from a mixture of state rebates, partnerships with manufacturers, charitable and corporate donations. They estimate it will have saved the 5,000 families they have worked with to date around £120 million over the lifetime of the installations, as it brings their utility bills down by 75-90%. They have trained 19,000 people since 2004, helping some of them to launch their careers in the solar business. Now they are spreading their wings, helping more communities across the US as well as down into Nicaragua.[32] What a great story.

INVESTOR-OWNED UTILITIES ENABLE ACCESS TO SOLAR RENEWABLES

Investor-owned utilities across the US are adding renewable energy generation to their portfolios, and some are doing it in a way that gets their customers participating in the process, rather than solely being 'at the end of the wire'.

One example of such a scheme is Bright Tucson Community Solar that is a scheme run by IOU Tucson Electric Corporation. Customers have the option to buy into solar without having to actually install a system on their property. The system, a ground mount solar array, was constructed and is owned by the utility. Customers are offered the chance to purchase blocks of solar energy upfront to offset their standard 'dirty' power. Whilst these blocks of power are currently more expensive than standard power, they are exempt from transmission surcharges and more importantly the rates are fixed for 20 years. So as the cost of electricity increases the pre-purchased blocks act as a hedge against increasing prices.[33]

It's a nice model that is certainly increasing access and penetration of solar and stabilising bills for participants in the process. But to me it is not really doing much to shift the status quo and create the 'prosumers' that become the agents of change in the system.

---

31 The Brewster Community Solar Garden® Project, www.brewstercommunitysolargarden.com
32 GRID Alternatives, www.gridalternatives.org
33 Bright Tucson Community Solar, www.tep.com/renewable/home/bright

RURAL CO-OP UTILITIES START TO EMBRACE RENEWABLES

Some of the rural electricity co-ops have been resistant to renewables because of worries it might push bills up. They have actually been lobbying against renewables. Other more enlightened co-ops have realised it has the potential to have the opposite effect and that actually investing and embracing renewables locally is good news for all involved.

One of the pioneers of this is the Farmers' Electric Co-operative, a 650-member co-operative in the town of Kalona, Iowa. Since 2008, it has become a national leader in installed solar watts per customer with a cumulative solar capacity of more than 1,800W per co-op member. It has deployed solar and other clean energy projects using a range of mechanisms. For those who self fund the installation of systems, there is a feed-in tariff or upfront rebate in place. For those who can't or don't want to install solar, they can buy solar generated power from a community solar 'garden' that has grown from an original 13.8-40kW and is continuously oversubscribed. Warren McKenna, the co-operative's general manager said, "The technology is coming, you're either going to embrace it, or a third party is going to come in and serve that load and you're going to lose that load."[34]

There is a huge opportunity for rural co-ops across the country to follow this great example of farmers, who are tied into fossil fuels for 50% of their supply, meaning that $1 million is leaving the community every year. The renewable energy they build slows that loss and keeps money and benefits locally as well as being better for the planet.

MUNICIPAL UTILITIES GO 100% FOSSIL FREE

Some of the municipal utilities are very active in driving the change to renewable energy. Not only do they generally provide cheaper electricity across the country than the investor-owned counterparts, some of them are also pioneering the change.

Thanks to its municipally owned utility, the largest city in Vermont, Burlington, is now 100% renewably powered. With a population of 42,000 the Burlington Electric Department (BED) has been on a mission to reduce energy consumption and increase renewable energy generation. Now its electricity is provided from wind, hydro and biomass including a 7.4MW hydro turbine.[35] The focus on local renewable energy generation actually came after a serious and long running energy efficiency programme. A programme that has meant

---

34 'Solar Success for Nonprofit Utilities – Farmers Electric Co-operative: A Small Rural Co-operative Becomes a Solar Leader', SEPA Case Study, www.solarelectricpower.org/media/230417/SEPA031-SolarOPs-Study_FEC_0914.pdf

35 'The US City That Has Decided to Go 100% Renewable', http://reneweconomy.com.au/2014/vermonts-largest-municipal-utility-goes-100-renewable-69042

annual electricity consumption in 2013 was about 5.3% lower than in 1989, even though the population grew by 8% over the same period. In 1990, people in Burlington voted to approve an €11.3 million bond to fund energy efficiency programmes that supported successful programme activities through to 2002. Since 2003, BED customers have been paying a small monthly charge that supports these programmes, meaning that almost $45 million has been invested in energy efficiency efforts sponsored by BED over the last 24 years. Nearly half of this expenditure was made by the customers. There is a big positive; not only are they now 100% renewable, energy efficiency investments save Burlington consumers about $10 million of retail electric costs annually.[36]

Pretty good work for a company owned by the people for the people, definitely travelling in the right direction. Of course there are some others taking this path, like the City of Palo Alto CA.[37] It makes me question why many more don't do this? Perhaps now the model is there they will.

CONSUMER CO-OPS GET SOLAR

There may not be lots of people using co-ops to build and own collective projects in the US, however there are lots of people coming together to form solar buying co-ops. All across the land people are setting up 'Solarize' or Solar Co-op schemes as some of them are called. Not-for-profit Solarize Washington has facilitated 600 home solar systems, and around $13 million investment from its local communities. By organising workshops in neighbourhoods, explaining the concept and then providing free assessments of people's solar potential, they bring communities together and help them access the best prices on solar in the area, by securing a bulk discount with one installation contractor.[38]

Anja Schoolman founded the Mount Pleasant Solar Co-operative in Washington DC with her teenage son Walter, and his friend, Diego. Because it was quite a lot of effort to set up solar they decided that whilst they were putting in the work they would make it possible for their whole community to do it at the same time:

> If we are going to do it, we are going to do the whole neighbourhood. Walter and Diego leafleted the neighbourhood to try and get people interested in going solar. Within a few weeks we had 50 signed up who were interested in going solar, and in 2007 we formed the Mount Pleasant Co-op. Together, we conducted energy audits and began trying to find

---

[36] 'Burlington Electric Department 2013 Energy Efficiency Annual Report', www.burlingtonelectric.com/sites/default/files/Documents/Energy_Eff/2013-energyefficiencyannualreport.pdf

[37] 'We Took the Leap – Palo Alto Electricity is Now 100% Carbon Neutral', www.cityofpaloalto.org/gov/depts/utl/residents/resources/pcm/carbon_neutral_portfolio.asp

[38] Solarize Washington – A Program of Northwest SEED, http://solarizewa.org

ways of conserving electricity. We also began surveying the roofs in the neighbourhood for installation of solar arrays.

By the end of 2010, they had completed 100 solar installations in the Mount Pleasant neighbourhood alone. The model Anja pioneered there has been adopted by many other communities in DC and the surrounding areas, growing into the DC SUN network of neighbourhood solar co-ops with 15 initiatives across the different regions helping over 700 individuals, churches and homes to go solar.[39] Anja has since gone on to found the Community Power Network that brings together grassroots, local, state, and national organisations working to spread the ideas and examples of community energy across the US.[40]

There are not just co-ops for homeowners; there are also co-ops of solar companies working together to leverage better prices. The Amicus Solar Co-op was founded in the autumn of 2011, 'when a small group of friendly solar installation companies decided it was time to form a special organization where independent businesses could support each other by sharing best practices and pooling their buying power'. Amicus is a purchasing co-operative that is jointly owned and democratically managed by the 32 members, operating across 23 states.[41] This is a great way to build a stronger industry and bring down prices to consumers at the same time.

SOLAR HOME FINANCE

I am aware that this piece on the renewable revolution in the US has been in a big part talking about different approaches for people to get access to solar PV, and there are many other technologies out there. However there is such a surge in the ways to access PV, I have been finding it hard not to share them all with you! What has been driving much of the uptake of solar are some of the big solar players like Solar City, Sunrun and Sungevity. They have taken solar mainstream by making it really easy for people to access the technology when they don't necessarily have the upfront cash to do so. Most solar companies in the US are offering solar with a finance package alongside it. So you can either take a 'solar lease' or a 'power purchase agreement'. Either way, you can get solar without putting any money down and experience savings from the moment the system goes live. In one case you lease the system often with an option to buy in time, and in the other you are simply locking in your electricity price and the savings for the lifetime of the installation. Like buying from the grid, but green, cheaper and at a fixed price. According to the Solar Energy Industries Association, "More than 90% of New Jersey's residential solar market has consisted of third-party owned systems since Q2 2013. In Q1 2014,

39   DC Solar United Neighborhoods (DC SUN), www.dcsun.org
40   The Community Power Network, www.communitypowernetwork.com
41   Amicus Solar Co-op, http://amicussolar.com

more than 50% of New York's distributed generation systems were third-party owned, and in California, Arizona and Colorado, 69-81% of installed distributed generation systems were third-party owned."[42] But these models once again leave the power and often the ownership somewhere else.

Understandably some people don't want to give their roof over to a third-party investor, and solar crowd funder Mosaic has sprung up to help people lend to each other to finance solar home systems. Peer-to-peer finance has come of age and people are more and more willing to lend each other money through these mechanisms. Mosaic's platform simplifies and de-risks the process for solar companies, individuals who lend and those who want to borrow to go solar. Bruce Ledesma explains one of the secrets to Mosaic's success:

> We are actually now taking a market share away from leasing because we created a loan product that is over a long period – 20 years. That allows for very low monthly repayments, certainly lower than your utility bill if you didn't go solar but also below every other loan product. It's about on par with the lease and so basically you have a choice; you can lease the system for 20 years and have to give it back or buy it for some undetermined value at the end of that period. Or you can buy the system, pay the same amount each month and own it after the fixed term. You then have about 15 years of clean energy after you've paid it off. So that's why it's starting to sell well.

Mosaic acts as a broker for the parties to make the transaction straightforward, and to reduce the risk and transactional cost for all involved.

COMMUNITIES CHOOSE CHEAPER AND CLEANER POWER

For me one of the most exciting stories from the US is 'Community Choice Aggregation' or CCA for short. It's one of the advances that has been driving the uptake of renewable energy and energy efficiency in some states. At the other end of the spectrum to those who are off grid, the CCA is all about helping people who are on grid to take back the power over their energy supplies, giving them choice in a monopolised system where there is none. CCAs are statutorily authorised retail electricity choice programmes administered by municipalities, which aggregate the demand of all customers within their jurisdiction. When they launch the scheme in an area, all the residents are automatically enrolled into the system, but have the right to opt out and remain with the existing supplier. The municipality is able to negotiate a favourable rate for the supply of electricity to the inhabitants and businesses, as well as being able to work on energy efficiency, and diversification of supply. Many aggregators also look to source their energy supply from renewable

---

42   Third-Party Solar Financing, www.seia.org/policy/finance-tax/third-party-financing

resources, which often in turn stimulates the development of new local renewable energy developments. As of 2014, CCA laws cover 25% of US annual electricity demand in seven states; with 1,300 municipalities involved it already serves one in 20 Americans.[43] As municipalities take control of energy supply on behalf of their residents it has multiple benefits: reducing consumer bills, providing new jobs and business locally and most importantly it starts to reduce greenhouse gas emissions.

The legislation to make this happen was first passed in Massachusetts and these early models passed on savings to consumers by using a broker to negotiate cheaper rates of electricity. Paul Fenn was the pioneer who came up with the idea originally and spent many years of hard work in making it happen. He started with the focus of acting on climate and emissions and concluded that giving people power over where their energy came from could be transformational. The first time he tried to table the idea as a Bill in 1995 to make CCA possible, it didn't go well: "The leadership were very unhappy with the Bill and the Senator who had hired me accepted a reassignment, and I was out of a job." After many years of trying, and repeatedly selling the concept to municipalities he managed to make it happen, and began rolling it out across many areas. "People often ask me how do you pass these Bills? And I say municipalities pass them because they have real political power."

Marin Clean Energy is a California CCA, started in 2008, that now serves 125,000 people with electricity. They have two tariffs, a light green 50% renewable energy tariff, and a deep green 100% renewable energy tariff. When compared to the standard offering from the local monopoly utility PG&E, which is currently offering a 22% renewable mix, the light green is slightly cheaper and the dark green slightly more expensive. I talked to Alex DiGiorgio of Marin CE about the CCA model:

It's a very replicable model and not something that many people know about. It defies conventional understanding about how the energy industry works and what is possible. The barriers to accessing renewable energy today are political and cultural not technological, they don't call electricity power by accident. In California most electricity users get their energy from a private regulated monopoly. Only 20% of the population are served by municipal utilities, they provide energy at a 20% cheaper rate than the monopolies. We are like a hybrid between the two in some ways. The municipal utilities own the hardware as well as supply the energy. As a CCA we don't own the hardware, we source the electricity, but we use the IOU to deliver it. We pay them for it, but we source the energy for our customers and the IOU still bills our clients. There really is no incentive for a monopoly to provide low rates, or more renewable energy, or even good

---

[43]   Local Power Inc. Community Choice Aggregation, www.localpower.com/
        CommunityChoiceAggregation.html

customer service. They are going to get paid either way.

In 2002, the Assembly Bill 117 was passed that made it possible for local governments to procure energy on behalf of constituents. So we are not a corporation or a company but a form of local government. Our board of directors is made up of the elected representatives of the cities they represent. They are the ones that approve our rates, they are the ones who approve our policies and programs. So CCA is really a way of democratizing the energy economy by giving customers a choice, but also because the folks who are making the decisions are elected by the people. Marin was the first to pursue this as a way of reducing greenhouse gas emissions and creating jobs.

When Marin CCA was setting up and starting to gain momentum the main utility in the area, PG&E, tried to stop it by funding a public ballot to the tune of $65 million to try and block them legally. Fortunately they lost and Marin went on to become both their partner and as well as their competitor.

"PG&E were very hostile and luckily we were able to weather the storm." Others were not so determined and therefore not successful. "PG&E and the other major utilities, they have a lot of political power, they have a lot of money and they haven't been afraid to use it. That's why people want competition."

Wherever a community forms one of these initiatives it becomes the default supplier. The population are mailed four or five times to let them know what is going on and they have the option to opt out. In Marin, people automatically get the 50% renewable energy option, which is a big improvement on the 22% renewable option offered by PG&E, but they are encouraged to 'opt up' to the deep green 100% renewable tariff. What they are working on right now is a 100% renewable, 100% local programme. It does cost more money, but the rate is locked in for 20 years so it is attractive for people.

The cost of our light green option is cheaper than PG&E, and MCE's. Customers are saving almost $6 million just this year (2014). Richmond City joined the MCE and their school district is saving £53,000 just this year. The examples go on and on, every city that is with us is saving tens of thousands of dollars. This scheme is not funded by tax dollars, it is all funded by money that would have otherwise gone to PG&E. We are just taking a share of what would have historically gone to them. It's the public option – or collective bargaining for energy and as a public agency we don't have to pay a dividend to shareholders.

The City of Richmond is very different demographically speaking than Marin County (it is the home of a huge Chevron Oil Refinery operation). It has a reputation of being a very challenged city. We launched with them in 2013 and more people in Richmond have chosen the 100% renewable tariff there than in any other district. This surprised a lot of people who think that communities who are less affluent are less interested in these issues, but it is actually the opposite.

I asked where does MCE go from here:

> We want to source as much of our energy from local projects as possible, to get away from the model of big generation plants in the middle of nowhere. We have 10 local projects at various stages of development. We have a feed-in tariff for projects developed in our area that gives clean energy projects a higher rate than market rates. We want to prioritise energy efficiency, and to create sustaining jobs. We also want to encourage other communities to start initiatives in their own area. This is accelerating the transition to low carbon – this year alone we have contracts for half a billion dollars of renewable energy – that's a pretty big injection into the economy. It's not about supply and demand. Whatever demographic people come from they often want more renewable energy, the issue is access, and that is what Marin and others are doing, opening up a market and giving the choice.

The originator of this groundbreaking idea, Paul Fenn, talks of the journey and the future:

> I am proud of what has happened. It amazes me that we have gone to one in 20 people supplied by CCAs. The Marin story where they are supplying 50% green power at rate parity is utterly persuasive. The next step is localisation and the use of finance, to establish some local generation and physical independence to the volatility of the fossil fuel industry.
>
> The work we have done in California has created 8,000 jobs and that gets the attention of mayors and officials, when the issue of climate change is not even talked about. The tide turned and we no longer had to make the argument about climate. In Illinois they went from nothing to about two thirds of the state served by CCAs in a year. In California people just didn't want to go there. But when the challenge by PG&E failed that changed and now we are working with a whole load of communities setting up their CCAs.

Paul also developed legislation for a bond issue to fund the building of locally owned renewable generation called the 'H Bond' to go hand in hand with the CCA legislation. The bond issue can be used to finance the construction of new renewable energy projects, and for those who invest the benefits are then passed back through the bills. Coupled with the CCA mechanism this has the potential to rapidly enable communities to move to high penetrations of renewable energy. "Using meter and transmission data to analyze the load, and local studies to analyze the renewable energy potential we worked out what is possible and the best way to deliver the optimal generation in a particular area. With the H Bond this opens up the equity position to the community but develops the projects in the optimal positions."

But just because it has worked in some places doesn't mean it is easy for a municipality to pull off a CCA and repower project:

> It is complex, and sometimes the simple thing to do is not the right thing to do. If you choose the easy thing you're going to get hurt; complexity is an opportunity not just a problem. I like to acknowledge the negative in this, to pursue the positive, as there is definitely a lot of both.
>
> Part of the problem with policy in the US and Europe is this notion that climate change can be averted simply by the addition of resources, but somebody is going to have to take a hit in the process. The fuel business, the transmission owners, if you reduce revenue on those systems they are going to take a hit. It's a bit like saying how do we reduce smoking without hurting the tobacco industries, it's not going to happen.

The CCA seems to be such a powerful mechanism because it doesn't shy away from confronting the monopoly business yet with democratic process at its heart. Once again it illustrates the power of collective action and the notion that some services like the provision of energy can be better achieved when profit is not the primary driver. Once the power is literally back in the people's hands choices can be made about how best to meet the energy needs. This in turn leads to analysis of what are the most cost-effective and stable solutions and of course local generation comes out on top. Perhaps the speeding roll out of CCAs across the US will mean they serve one in four people in the next 10 years rather than the 1 in 20 they have achieved in the first 20 years. Many of the existing schemes already have renewable energy master plans for provision of local power in their area, and I am sure the coming years will see many succeed in producing significant amounts of the energy their communities need from within the communities themselves. It is truly transformational and a testament to the power of individual determination leading to collective action. An illustration of the power of good ideas to spread.

There is such a diversity of responses across the US and so many good examples of local action to adopt clean energy. However, there are many challenges faced by all those involved in the renewable roll out. The list of challenges to 'net metering' and distributed renewables by the incumbent utilities is pretty huge. Just one example, "the state House in Oklahoma ... passed a bill that would levy a new fee on those who generate their own energy through solar equipment or wind turbines on their property".[44] The headline sums it up nicely: 'The World's Dumbest Idea: Taxing Solar Energy'. In another

---

[44] 'The World's Dumbest Idea: Taxing Solar Energy', http://news.yahoo.com/worlds-dumbest-idea-taxing-solar-energy-111300623.html

case Dominion Virginia Power, the largest utility in Virginia, successfully blocked a solar net metering bill from moving forward. In fact Dominion lobbied for 'substitute language that would give the utility the exclusive right to build and own community systems and sell the power to the customers' thereby forcing multi-family residents to purchase solar only from Dominion.[45] This seems to be a repeating pattern in many states, as IOUs wake up to the game-changing potential of distributed generation. Perhaps inevitable when faced with the change that is coming. Utilities "do realize that distributed solar is a mortal threat to their business," according to David Crane, president and chief executive of NRG Energy. He continued: "Traditional transmission and distribution utilities will have to deal with distributed solar power, and it won't be a pretty fight."[46]

FULL CIRCLE

The one thing that is needed when facing these challenges is community support and collective action. Miles from these fights over power, infrastructure and big business, the battle is being fought and the revolution is underway in another community. Henry Red Cloud is a solar warrior and founder of Lakota Solar Enterprises.[47] He is from the Oglala Sioux Tribe from South Dakota, fifth generation descendant from Chief Red Cloud, and has been a pioneer in renewable energy in his community for the last 15 years.

The community in Pine Ridge is amongst the poorest communities in North America. Poverty is widespread and access to energy is a huge problem for many. People are living in poor quality homes that are hard to heat, often in very low temperatures. Henry set up his company to address these issues one house at a time. He began manufacturing and installing solar air heating and other renewable energy systems, such as PV and small wind, across the reservation. He has trained over 400 people in renewable energy – mostly from other tribal groups – and created over 160 jobs in an area where unemployment is endemic. His work with solar air heating drops people's heating bills by a third, using solar air collectors that have been designed and manufactured locally.

> The reason why I do these things is to create a better way – a better future, a better economic justice, as well as doing my warrior deed here in the 21st century. Our culture tells us that we are warriors, and we are always going out getting resources and bringing them back for a better quality of

---

45 'Dominion Virginia Power Blocks Net Metering Bill', www.cleantechnica.com/2014/02/10/dominion-virginia-power-blocks-net-metering-bill/

46 'Utilities Facing a 'Mortal Threat' From Solar', http://blogs.wsj.com/corporate-intelligence/2013/03/22/utility-boss-faces-mortal-threat-from-solar/

47 Lakota Solar Enterprises, www.lakotasolarenterprises.com

life for the tribes, so basically that is what I am doing. Creating economic opportunities and using a new way to honor the old ways.

Renewable technologies fall into our way of life, we as native people have always been embracing renewable energies. It's in our language, our dance, our song, and our ceremony. We are taking something that we understand and implementing it in a 21st century way. It's all about healing oneself, and then the family, then the community, then the tribe.

We are rebuilding through our culture, we are rebuilding through renewable energy, we are moving forwards as nations. I am sharing all of this stuff that I know about renewable energy. Sharing this to create hope, to create jobs. Native Americans' culture is based around the Earth – we are the caretakers of mother Earth. We are doing our part in a little way. If we all do our part we can make it a better place.

Native people – they think in generations, they base their plan and strategy on how it is going to effect four to seven generations. That's our mentality and way of thinking, the Lakota people. So that's where I am at, creating a better way for the future people who come. At the start I was the only native business in renewable energy in the whole US – I could have taken advantage of it, but that wasn't my calling. My calling was to share the work, and get people saving money and creating a better lifestyle for themselves, their family, their community and their tribe.

CHAPTER 11

AFRICA

"Sub-Saharan Africa is rich in energy resources, but very poor in energy supply. The picture varies widely across the region, but, in sub-Saharan Africa as a whole, only 290 million out of 915 million people have access to electricity and the total number without access is rising."[48]

According to an International Energy Agency estimate the annual residential electricity consumption in sub-Saharan Africa is the same as electricity consumption in New York State, when South Africa is excluded. The 19.5 million people in New York State, US, consume the same amount of electricity as the 791 million people of sub-Saharan Africa.

In many countries in the region wood fuel is still the primary energy resource used for cooking – in fact it is the primary source of energy. Collection of this wood and production of charcoal is often linked to deforestation and its use as a fuel to ill health as people cook on open fires inside their homes and inhale the smoke.

For those of us living in industrialised nations where flicking the switch as it gets dark, cooking on gas and the heating coming on with a timer are normal, the life experienced by those hundreds of millions of people is a stark contrast. In fact I find it hard to imagine what life would really be like without access to energy.

In 2001 I was a volunteer in a rural Tanzanian village and experienced life in a house with no infrastructure and no lighting. You are confronted with it – it's pretty abrupt, people have no running water, they have to get firewood for their cooking, batteries for their radio. My batteries rapidly ran out and then I would have to travel two or three hours to the nearest town to get new ones. I was thinking I would have lots of time to do stuff in the evenings, like read. As a volunteer we were given a kerosene light to use, it was a pain, the lighting was poor, and it gave me a headache so in the end I just went to bed. This was what everyone else was doing. The village goes to sleep at 6.30pm. Walking away from that village I wondered how this would ever change, next year or even in 10 or 50 years. This wouldn't change without some sort of intervention.

---

[48] *Africa Energy Outlook: A Focus on Energy Prospects in Sub-Saharan Africa.* 2014, International Energy Agency.

John Keane has since gone on to become the CEO of Sunny Money, delivering over a million solar lights to rural communities in five African countries. That's quite some intervention.

The last few years has seen great innovation in addressing this problem, with a large number of companies springing up to try and address the issue of access to energy across rural Africa, both the lack of electricity, and also the use of wood for cooking. There is a rush of entrepreneurial activity in the region around energy with local people getting started as well as international entrepreneurs getting involved. Energy supply in many parts of Africa is intermittent and more crucially the grid network does not reach the majority of the population. A bit like the US in the 1930s, the urban centres in some areas have electricity grids and energy infrastructure, but rural areas almost certainly don't. It is also quite likely that they never will if they are reliant on the central power station model to reach them.

If people have access to light it has been by burning kerosene, candles, open fire or torches. Kerosene is grim stuff: expensive, polluting and dangerous. Where kerosene is available it is more expensive in rural areas, and people tend to buy it in very small quantities, less than a litre. It is often sold in re-used soda bottles so it is quite common for children to mistakenly drink it. "Across the African continent, an estimated 53% of off-grid households use kerosene-based lighting and spend $4.4 billion a year on the fuel. Using poor quality lighting from kerosene lamps increases the risk of household fires, personal injuries such as burns and kerosene poisoning and, in the long term, the risk of chronic respiratory diseases due to harmful emissions."[49]

Not only does a kerosene light give off toxic fumes, kicking the light over is a real possibility and a real hazard when you live in a home made from flammable materials. To put it in context, a study for the World Health Organisation on low-income households in South Africa cited kerosene as the second highest cause of toddler death in the community.[50] Horrendous stuff.

If that isn't enough, kerosene use is also financially costly, with up to 20% of household expenditure going on kerosene for access to light. The solar light – a $10 product combining an LED light, a battery and solar panel – offers an antidote to the kerosene light problem and a first step for people to reduce its harmful effects.

John Keane explains:

> The concept and vision of Solaraid[51] (the charity that is behind Sunny Money) was that worldwide everyone should have access to clean energy.

[49] Esper, H., London, T. and Kanchwala, Y. (2013). *Access to Clean Lighting and its Impact on Children: An Exploration of SolarAid's SunnyMoney*. Child Impact Case Study No. 4. Ann Arbor: The William Davidson Institute. http://wdi.umich.edu/research/bop/projects/field-based-projects/Child%20Impact%20Case%20Study%204%20-%20Access%20to%20Clean%20Lighting%20-%20SunnyMoney.pdf

[50] www.who.int/bulletin/volumes/87/9/08-057505-table-T2.html

[51] www.solar-aid.org

The mission was to eradicate the kerosene light by 2020. We tried lots of different ways of getting the solar light out there; it took a while to get to scale. This may not be a bad thing as the products have also been evolving and increasing in reliability. People were not designing lights for rural Africa – this whole sector didn't exist a few years ago. The nice thing now is that there is a growing market. The lights are being designed specifically for use by people living in rural Africa, India and Bangladesh. Now everything is better, the quality of the build and the design of the product. The developments in batteries and LED technologies mean also that you get a lot more for your $10.

Our story is of recognising that need. If everyone is using fire at night for lighting, people are getting burned, people are dying, and children are fairly frequently ingesting kerosene. That's why we wanted to reach scale. The solar lighting products may have been in camping shops in the UK for 10 years but they haven't been in African villages. We asked, how do we change that? We focused on last mile distribution so that the products that are suited to rural African households are made available in those areas and people actually see them. There is a latent need. Our main sales channel has been through schools networks, working with teachers, as people trust teachers. It's an aligned mission in many ways – if a school child has access to a solar light, that means they can study. It's a win-win.

That's why we have seen our sales go from 10,000 in 2009 to over 615,000 last year. The constraints have not necessarily been people keen to buy the products, although there are still some who cannot afford them. It is all the other stuff that is stopping the growth – in an ideal world we would be in 30 countries by now. Across five countries we have found there are networks that are in every village across Africa, and accessing these networks is a way to get solar lights out across the continent.

A solar light changes the way you operate overnight. Once you have got it you don't have to wait for an electricity grid that may never come. It doesn't replace the electricity grid but it's a darn sight better than a 19th century kerosene light.

We are part of a growing sector. Access to finance is an issue. There are still people who can't buy the $10 product. We put in a pay as you go chip in the lights, which enables people to pay in installments. They spend one or two dollars and then they have got the light, and they pay the balance in installments over up to 12 weeks. Then the light is theirs. The payments mimic kerosene purchase, but at the end they own the light.

We are shooting for one million lights this year, and there is increased competition, which is good if the quality is good. The market does have the challenge of poor quality products being sold that won't last. It's a whole other battle. But the growth will continue in the sector. We do plan to go into more countries – another eight by 2017. We want to share a lot of our workbook as we have this mission for the end of the decade and we know we can't do that alone.

The solar light is the starting point for people to access energy, and Sunny Money and the other players in the sector offer a range of products that allow people to do more, like adding phone charging. It is the start of the 'energy escalator'.

In the last 15 years I have seen three things that have been transform- ational. The first was the mobile phone, which has changed the way we communicate and do other things like banking. The second was affordable motorbikes; they have mobilised rural Africa. The third is the solar light. All these things can happen really quickly. It's not slow growth, it's transformational rapid stuff.

The mobile phone effect has inspired the team at Kenyan company M-Kopa to help people access energy using mobile technology as an enabler. M-Kopa sells solar home systems, a step up the 'energy escalator' from the solar lights across Kenya, Uganda and Tanzania. The systems provide lights, phone charging, a radio and a solar panel all linked to a GSM mobile unit. The embedded mobile system means that people pay via mobile for their system over the course of a year once the initial deposit has been taken. This gets round the issue of upfront investment, and Kenya has one of the highest uses of mobile payment systems anywhere in the world.

However mobile payment systems don't reach everywhere and there is demand for systems that provide more energy than lighting and phone charging. BBOXX is a company that has focused on meeting this need, delivering its products through 30 shops in Kenya, Rwanda and Uganda as well as in over 30 other countries. Mansoor Hamayun, BBOXX co-founder and CEO, was studying electrical engineering and was starting to question why he was there when so many people around the world did not even have access to electricity. With a few friends, first he set up a charity to try and address the issue. Following on from their successes with the charity they established BBOXXto develop the work further.

I had seen lots of the solar products out in the field, and most of them were delivering photons rather than electrons, and the idea stemmed from that. We wanted to create a range of products that could deliver the on-grid experience in off-grid areas. We saw a huge potential to work with the already successful businesses in the areas. We just crossed 50,000 installations a few months ago.

BBOXX is selling systems that can power a house, and are often being purchased to replace or back up diesel generation. Some systems are grid tied, which means that the intermittency problems often experienced on the grid are no longer an issue for the BBOXX user.

All of the systems they deploy are monitored remotely, and can be managed from an office elsewhere. This monitoring means that they can analyse how people are using the systems, but also spot problems easily.

> In every country where we work there are three problems that are inherent: physical access to the products and services – the distribution channels don't really exist, information and awareness – people are not aware of solar, and the final one is financing. The solutions to these problems are different in different areas. We have been focusing less on the rural customers. You can connect our products to a diesel generator and the grid easily so people like that. The payback can be less than 12 months for some systems. We are giving people what they need, but we are also slowly moving to giving them what they want – and they are two different things. The people we are working with are not beneficiaries, they are customers.

People across many African countries are making a clear choice: renewables ahead of fossil fuels, and they are making it on a cost, as well as ease of use, basis. Once people understand about renewables and particularly solar in these communities – why would they use anything else?

Village Infrastructure has taken this thinking a step further by not just coming up with a technology, but also a business model to empower local entrepreneurs as well. Co-founder, Stewart Craine, is a trailblazer in the area of rural electrification, founding Barefoot Power and spending six years as its CEO. Barefoot Power products have provided light to two million people in 22 countries. Not stopping with this considerable success his next project, Village Infrastructure, has built a new model to empower energy entrepreneurs in rural Africa. The business is focused on delivering 'mini-grid' systems – combining solar and batteries and a mini-grid network to provide people with the energy they need on a village wide basis. Basically building small-distributed utilities where they are needed most.

> The idea is to create a power company where the assets are put into the villages at credit with very low upfront costs and to attract repayment over time. Extending credit to the people who need it most, and pushing out repayment periods from six or 12 months to one, two, three, maybe even 10 year loans. This is done to amortise the cost of assets over time and make them affordable for everyone, to get deep market penetration and ultimately deliver energy access for all. I don't think the cash sale model is going to get energy access to everybody because you can't sell a mini-grid for cash.
>
> We look to share 10% of the revenue with village based energy agents so that we are creating a job in every village we go to. The agent manages the

power station in the village and so builds up skills hopefully even if it is on pen and paper only, but if they can manage pen and paper there is a good chance they can manage the technology as well.

Our staff are the equivalent of franchise managers; they manage a cluster of 50 or 100 local energy entrepreneurs running the local power stations in the villages with a 20-30 kilometre radius. We are using a structure that has been tried and proven, similar to a group-lending leader in microfinance and you will have a team of about 15 women that actually collects money. On a village basis 1-3kW supplies an entire village in LED lights.

To put this in context, a single home in Europe may have more solar installed on it than that – a 3kW system is pretty standard for one house. Village infrastructure has focused on mini-grid systems as they are working on the premise that supplying light is just the start. To really transform rural lives, access to things like water pumping and milling will be crucial. Being able to provide small agricultural processing machines can transform villages' ability to preserve their crops and generate cash from them as well. Putting a micro-grid in place means that villages would potentially have enough power to run a small milling device, rather than having to take their crops to a larger village and paying the associated costs. Simple changes can make all the difference to a community's wellbeing.

The grid infrastructure in many countries, as well as not reaching the majority of the population, is under pressure serving the users it does reach. As such blackouts and 'load shedding' are common in many areas; it is normal for businesses to have standby diesel generators to back up their supply when the power does go off. Diesel generators are expensive to buy, run and maintain on top of electricity costs that are also often high. This means there is a huge opportunity to deploy renewable energy directly to reduce intermittency and costs, and in many situations across Africa it is cheaper than power from fossil fuels. A recent study on the state of solar PV globally on behalf of the EU put it like this: "Solar PV electricity is now the cheapest electricity option for more than one-third of the African population."[52] Personally I think it is probably higher.

As such, all over the continent solar and other renewable energy is being deployed at various scales. It's not just the off-grid market that is turning to renewable energy. On grid there are renewable energy developments in pretty much every country on the continent; to give you a brief flavour of what's underway:

In Ghana, the current energy generating capacity is 2,100MW. Construction is underway on the largest single utility-scale solar PV park in Africa

---

52   Jäger-Waldau, Arnulf (2014). *PV Status Report 2014*. Joint Research Centre, Institute for Energy and Transport.

at 155MW.[53] There are also other large-scale solar plants being built in the country and one plan to build an additional 600MW of solar, which would add 28% to the total production capacity of the country.[54] There are also 200MW of wind turbines being constructed that will be equivalent to 10% of the country's total generation.[55] Although more than 60% of the population has access to the electricity grid, there are still many who won't have for the foreseeable future, so there are also schemes rolling out standalone solar powered systems to schools and homes across the country.[56]

In Rwanda, only 22% of the population has access to electricity. The total installed generation capacity in the country is just 155MW. Most of this generation is from hydroelectric and diesel, however there is already a 8.75MW solar farm constructed, with more under construction. Once again the speed of roll out of the grid is not fast enough and generation cannot keep up with demand so an ambitious solar home programme is underway to deploy 50,000 systems across one area of the country in the next two years.[57] Others are deploying micro-grid systems in the country, electrifying whole villages at a time. Mesh Power, a start-up from London, already has their seventh mini-grid installed.[58]

In Morocco, they are planning to get 42% of their energy needs from renewable energy by 2020, and to get them on the path they have just commissioned Africa's largest wind farm. The 301MW Tarfaya Wind Farm situated on the Moroccan Southern Atlantic Coast will provide power for 1.5 million homes.[59] They have just completed tendering for the construction and operation of a 350MW solar thermal power station, which once completed will provide 18% of the country's energy needs.[60] Impressive stuff.

I had the pleasure of taking a short trip to Ethiopia in 2013 with the then UK Energy Minister, Greg Barker, and was struck by how central the plan for sustainable development was in all government ministries that we visited – they are aiming to be a carbon neutral country by 2025. From the Prime Minister, there was a clear focus on meeting both the environmental as well as the social challenges that the country faces, heartening stuff with buy-in from

53   'Ghana to house Africa's Largest Utility Scale Solar Power Plant', www.pennenergy.com/articles/
       pennenergy/2014/03/ghana-to-house-africa-s-largest-utility-scale-solar-power-plant.html
54   'Huge Solar PV Power & Manufacturing Projects Headed To Ghana', www.cleantechnica.
       com/2014/03/03/huge-solar-pv-power-manufacturing-projects-headed-ghana/
55   'Ghana to Get 200MW Wind Power by 2016', www.citifmonline.com/2014/10/29/ghana-to-get-
       200mw-wind-power-by-2016-kofi-buah/#sthash.FX2sdOw0.dpbs
56   'Ghana Spurs PV Development Across Country', www.pv-magazine.com/news/details/beitrag/ghana-
       spurs-pv-development-across-country_100017701/#axzz3R71GTfLT
57   'Rwanda Turns to Small Solar Units to Expand Power Access', http://af.reuters.com/article/
       rwandaNews/idAFL6N0U10U220141217?pageNumber=2&virtualBrandChannel=0
58   A Fortnight in Rwanda, www.meshpower.co.uk/blog/
59   'Africa's Largest Wind Energy Project Commissioned In Morocco', www.cleantechnica.
       com/2014/12/31/africas-largest-wind-energy-project-commissioned-morocco/
60   'Morocco Moves Forward With 350MW Solar Thermal Power Bids', www.cleantechnica.
       com/2015/01/19/morocco-moves-forward-350-mw-solar-thermal-power-bids/

across the leadership. Deployment of renewables across the country is already happening and they have a stated ambition to become a green powerhouse in Africa. Hydropower already accounts for the majority of the power on the grid in Ethiopia, and more hydro projects are underway. However only 20% of the population has access to the grid, so expanding that access is a major priority. According to the Ministry of Water and Energy, at the end of 2013, the country had about 826,000 solar home systems as well as solar systems for 345 rural health centres and 190 schools. On top of that a further 300MW of large-scale solar has also been approved.[61] Wind power has begun to be exploited, with more being developed all the time. In 2013 a project was erected north of Addis Ababa that was the largest wind project in Africa at the time at 120MW.[62] Africa's largest geothermal power station, a 1,000MW project, is currently under construction in Ethiopia, to add to other geothermal plants already operating in the country.[63] I had the joy of experiencing the geothermal resource first hand – as the pool in the hotel I stayed at in Addis Ababa was actually heated by geothermal – a real treat!

In fact in 2014 across sub-Saharan Africa, more renewable energy generation plants were constructed than in the 13-year period before. Around 1.8GW of new renewable energy generation came online in 2014 as renewable energy comes to life across the continent.[64] First steps in the unfolding energy revolution.

It is not only access to light and electricity that is beginning to change. The use of wood and charcoal for cooking has started to be transformed as well. Cooking on open fires or even with charcoal in small homes produces smoke and fumes that are harmful to health, and requires a lot of fuel. The wood and charcoal is often coming from areas that end up being deforested or degraded in the process. Areas of national parks are illegally cut to serve this need. However the basic open fire that is being used is really wasteful in terms of its ability to transform wood or charcoal into useful heat. The 'efficient' cookstove, or 'Jiko' as they are known in some areas, can start to change this by reducing the consumption of fuel by half. Since the 1980s businesses and NGOs across the continent have been manufacturing and supplying these simple stoves that are generally constructed from locally available materials.

The reduced need for wood or charcoal saves the users money and slows the speed of deforestation, and above all prevents huge health issues. As the Global Alliance for Clean Cookstoves explains:

---

[61] Jäger-Waldau, Arnulf (2014). *PV Status Report 2014*. Joint Research Centre, Institute for Energy and Transport.

[62] 'Ethiopia Opens Africa's Largest Wind Farm to Boost Power Production', www.reuters.com/article/2013/10/26/us-ethiopia-electricity-windpower-idUSBRE99P04Z20131026

[63] 'Ethiopia to Build Africa's Largest Geothermal Plant', www.oilprice.com/Alternative-Energy/Geothermal-Energy/Ethiopia-to-Build-Africas-Largest-Geothermal-Plant.html

[64] 'Sub-Saharan Africa Commissions 1.8GW RE Capacity This Year', www.pv-magazine.com/news/details/beitrag/sub-saharan-africa-commissions-18-gw-re-capacity-this-year_100016179/#ixzz3RCgm8yce

Daily exposure to toxic smoke from traditional cooking practices is one of the world's biggest – but least well-known – killers. Penetrating deep into the lungs of its victims, this acrid smoke causes a range of deadly chronic and acute health effects such as child pneumonia, lung cancer, chronic obstructive pulmonary disease and heart disease, as well as low birth-weights in children born to mothers whose pregnancies are spent breathing toxic fumes from traditional cookstoves. The evidence is robust and compelling. Exposure to these toxic fumes is greatest among women and young children, who spend a disproportionate amount of time near open fires or traditional cookstoves.[65]

The World Health Organisation (WHO) estimates that globally, over four million people die prematurely from illness attributable to the household air pollution from cooking with solid fuels. It also states that more than 50% of premature deaths among children under five are due to pneumonia caused by soot inhaled from household air pollution.[66] Awful statistics. Clean cookstoves can have a huge impact on reducing these terrible impacts.

Mark Simpson is founder and MD of C02balance, a company that has specialised in delivering clean cookstoves to communities that need them across a number of African countries:

One of the biggest impacts that people across any African country have is cooking on open fires. This has massive health issues because of the particulates, smoke and the carbon monoxide. It is also massively inefficient in terms of resources. Ten years ago when we started there was not much on the market, so we developed our own stove that was brick built, made from local materials. We employed a stove engineer to redesign the stove so it had a better fit with local culture and practices. We sold the scheme as a carbon offset and that helped to fund the stoves and expanded it through Kenya. Then we looked at rolling it out in other countries: Ethiopia, Tanzania, Ghana.

Ten years on and they have delivered nearly 100,000 stoves into rural communities across Africa and have used carbon offsetting to help make the stoves affordable to the poorest communities.

The impact on people's lives is huge. Children and older people's health are massively improved. One of the biggest impacts is that the women of the family spend less time collecting wood. The mother in a household

[65] 'Global Alliance for Clean Cookstoves – Health Impacts', www.cleancookstoves.org/impact-areas/health/
[66] 'Household Air Pollution and Health – World Health Organisation', www.who.int/mediacentre/factsheets/fs292/en

could be spending 12 or 16 hours collecting wood a week. Once they have a stove this will drop to four or five. Lots of the communities we have worked in have set up small co-operatives for bag making or basket weaving in the time that they have gained. It has huge economic benefits. So the demand is growing as one village gets some stoves then another realizes the opportunity.

For me opportunity is a key word when considering what is unfolding across Africa today. The opportunity for communities and individuals to create change through clean and renewable technology is massive. The opportunity for new businesses and jobs is huge. The opportunity to improve health conditions, reduce environmental impact and increase wellbeing of communities is real and achievable. Delivering a grid connection to many people across Africa is unlikely to happen any time soon. However the distributed models described above mean that access to energy for those people is not only technically possible, but also cost effective. As people in Africa leapfrogged the standard telephone system and went straight onto mobile, it seems inevitable that a large number will go straight to distributed and renewable energy for their energy provision. In the process they will be bypassing the central generation and distribution system that is common in the industrialised nations. The opportunity is for the new distributed model to be pioneered here. Many people are already creating the examples for others to follow in response to an urgent need – perhaps Africa is the incubator of the next pioneers of the energy revolution. Let us hope so.

CHAPTER 12

BANGLADESH

Bangladesh is one of the world's most densely populated countries with 150 million residents. Criss-crossed with rivers and lying on a large delta, it is also very low lying. It's a country often known for extreme poverty and devastating floods. Bangladesh is already in deep water in terms of experiencing the effects of climate change – some have called it the climate change 'ground zero'. It is already losing land and the IPCC has estimated that the country will lose 17% of its land area and 30% of its food production by 2030. It seems now that these figures are probably an underestimation. This process is already underway as powerful cyclones surge up the Bay of Bengal destroying homes, land and livelihoods. Lots of the population rely on agriculture for their income and sustenance. Not only is land being lost to rising waters, the land that is left is becoming more saline so that crops are less viable.

Many of the population have no access to electricity, with nearly 40% of the population off grid, relying on biomass and waste for cooking and heating. For those who do have access to the electrical grid, this is no guarantee of consistent power. Shortages in the supply of natural gas have hit the electricity production sector, meaning that power outages are frequent. Natural gas made up over 50% of the country's primary energy consumption in 2012, with another 25% coming from waste and traditional biomass fuels.[67]

But in this country with so many challenges, a new story has unfolded, a story about reaching people in the areas where the grid doesn't reach. Bangladesh is a world leader in distributed small-scale renewable energy. Renewable energy and particularly distributed solar is already widely used in the country. There are now over three million solar home systems in operation, providing families with clean and cheap power, and reaching communities that would otherwise have no access to energy. This substantial roll out of solar technology touches many millions of lives, and started as early as 2002 with 47 organisations taking part in a programme backed by the World Bank as well as other international aid organisations, philanthropic funders and international banks.

One of the pioneering organisations leading this was Grameen Shakti, established in 1996 by Nobel Peace Prize winner, Professor Muhammad

---

[67] EIA Country Analysis – Bangladesh, www.eia.gov/countries/country-data.cfm

Yunus, as a spin-off from the highly successful and pioneering micro credit bank, the Grameen Bank. Grameen Shakti, which literally means 'village power', is a social enterprise with a social and financial mission. Its aim is the successful dissemination of affordable, renewable energy technologies to rural Bangladeshis. Following a trial to establish if home solar systems were viable in the country in the mid-1990s, it slowly scaled up its activities until by the end of March 2015 it had deployed nearly 1.6 million systems. That's half the entire systems installed in the country and an amazing achievement.

It has worked with international funds and donors to be able to offer all the systems it sells with a finance package that makes them affordable to rural households. Sometimes they have been able to reduce the price of the systems with grant funding from the same sources. This has been key to enabling access, and the company's approach to providing funding is drawn directly from the Grameen Bank model that had been so successful in the country and has once again worked exceptionally well.

The company's international profile meant that in time they were able to access funds from international development agencies, channelled through the country's own Infrastructure Development Company Limited (IDCOL).

I spoke to one of the Directors of Grameen Shakti, Mr Hassan, about what they have achieved: "We have distributed these systems to people who don't have regular monthly incomes, as their income is agriculturally based. Sometimes people struggle to meet their monthly payment because of that, but we can be flexible with it. So far 92-95% of the loans we make are repaid." Over 600,000 loans have been repaid in full already. That's 600,000 households who have achieved energy independence.

Over the years Grameen Shakti has also expanded its product range to add clean cookstoves and small-scale biogas plants. It has sold over 900,000 clean cookstoves and constructed over 30,000 biogas plants. So far they estimate their work has touched the lives of nearly 18 million people in the country. I find it staggering to think that one company on a mission could have such a profound effect on the lives of over 10% of the population.

The solar systems they are providing are starting at as little as 10Wp to provide a home with two lights and go up to 120Wp. They offer a range of funding options to ensure that the systems are affordable for more people, spreading the cost of purchase over potentially three years.

In the process of deploying this vast number of systems they have built a large network across the country, based out of 1,500 offices with an army of 11,000 employees. Mr Hassan explained: "We have around 35 'Grameen' technology centres around the country. In the centres we train underprivileged groups and particularly women." To date they have trained over 20,000 technicians and nearly a million users in the technology.[68] Many of the people

[68] Grameen Shakti – Programs at a Glance, www.gshakti.org/index.php?option=com_
content&view=category&layout=blog&id=54&Itemid=78

they are training are women, who go on to work for the company and free-lance installing and maintaining solar systems. Hassan continues: "They can work as micro entrepreneurs. We also have a micro utility model – if people purchase a package that has say four lights, they maybe have two lights more than they need, they can easily rent two lights to a neighbour and earn some money on the lights." So not only are the company providing access to energy, they are providing significant employment, training and entrepreneurial opportunities for many people across the country. Increasing the wellbeing of the community on many levels.

Mr Hassan continued:

> Bangladesh has a very large market for solar home systems. We are looking at micro-grids for people who cannot afford solar home systems, maybe 5kWp systems supplying many homes at once. It's a new phase for us in Bangladesh, we have only one built at present, but we hope to have many more constructed in the coming years. Solar irrigation is another focus for us in the future.

It seems the speed of roll out of these systems is set to rise. Across the programme some 50,000 systems are being installed every month, and the level of ambition doesn't seem to be slowing. IDCOL has a target to finance six million solar home systems by 2017; this will have an estimated generation capacity of 220MW of electricity. This would be doubling the market from where they are in 2015, but happening in two years and not the 13 years it took to get the first three million systems installed.[69] It just illustrates the staggering speed of change that is occurring as solar and other renewables come down rapidly.

One of the early pioneers of this area, Dipal C. Barua, a former Managing Director of Grameen Shakti, is quoted in an article on Reuters saying: "My dream is to empower 75 million Bangladeshis through renewable energy by 2020 and make Bangladesh the first comprehensive solar nation of the world."[70] Despite all of the challenges they face as a nation, they have made solid steps to making that vision a reality. The race is on to become the first solar nation, and Bangladesh has a good headstart.

---

[69] Infrastructure Development Company Limited (IDCOL), Solar Home Systems, http://idcol.org/home/solar

[70] 'Bangladesh Aims to Be World's "First Solar Nation"', http://in.reuters.com/article/2015/01/25/bangladesh-solar-idINKBN0KY0O220150125

CHAPTER 13

CHILE

Chile is an amazing country, only 350km wide yet over 4,000km long. In the south are lakes and mountains, volcanoes, glaciers and fjords. To reach some areas in the far south of the country the only way to travel is either by boat or plane, as no roads cross huge areas. In contrast with this, in the north is the vast Atacama Desert, supposedly the driest and sunniest place on Earth. I had the pleasure of travelling the length of the country when my parents spent three years living there, and it is an amazing place.

Chile is rich in natural resources and natural beauty, but it is not endowed with a source of fossil fuels, so it is reliant on importing them from neighbouring countries and further afield. Chile has one of the highest electricity prices in Latin America, partly because 97% of the fossil fuel energy it consumed was imported in 2012.[71] One third of its electrical generation comes from hydroelectricity, and most of this is connected to the southern grid (there are two grids operating in this vast country). However five years of drought has meant that it is even more dependent on fossil fuel generation, and that pushes prices up.

Unsurprisingly, Chile is experiencing a boom in renewable energy deployment as people and businesses throughout the country are waking up to its benefits. Between January and July 2014, the country added 600MW of renewable energy capacity to its grids, more than twice as much as in the whole of 2013. Renewables now account for nearly 9% of Chile's installed capacity.[72]

One of the sectors that is driving this uptake is the mining industry in the north of the country. The vast copper mines in the Atacama Desert consume 39% of the country's power, and are powered with a mix of imported gas, coal-fired power stations and diesel. This is not only expensive but it is also dirty. These mines are turning to renewable energy to power their operations.

As such, solar has arrived and is starting to be built on a pretty large scale. There is no feed-in tariff or similar scheme in the country, so some of the systems are being built and connected directly to the mining operations. One development called Project Salvador was constructed by US company SunPower and connected to the grid in the Atacama Desert. At 70MW this is the second largest PV plant to go online in Chile. It is the largest merchant

---

[71] EIA Country Analysis Note – Chile, www.eia.gov/countries/country-data.cfm

[72] 'Energy in Chile – Winds of Change', www.economist.com/blogs/americasview/2014/09/energy-chile

solar plant to be completed in the world. What this means is that the plant has no power purchase agreement. It therefore sells the electricity it generates on the spot market competing with all other forms of generation. This is a world first for solar.[73]

Wind energy is being deployed for the same purpose, again often directly connecting to the mines. The largest scheme to date sits on a coastal hilltop 400km north of Santiago. With a peak output of 115MW, the output from the 50 turbines will feed energy directly to Los Pelambres, a nearby mine. It will provide around 20% of the mine's electricity.

Wind and solar are not the only options in the country. Chile's 137 volcanoes are being evaluated as potential sites for geothermal power plants.[74] As in many countries there are quite a number of obstacles for the deployment of renewable energy on the grid and the more complex projects like geothermal are harder to develop.

Chile, however, is one of the most active countries in the world for renewable deployment. In northern Chile, solar is cost competitive with conventional energy and developers are queuing up make the most of it.[75] Individuals and communities have yet to really start waking up to the benefits of renewables, but a net metering law and the growth of renewables across the country means that it is only a matter of time before they do.

---

[73] 'World's Largest Merchant Solar Project Goes Online in Chile', www.pv-magazine.com/news/details/beitrag/worlds-largest-merchant-solar-project-goes-online-in-chile_100017058/#ixzz3T3hExqex

[74] 'Chile Top Renewables Market on Sunny Desert, Windy Shores', www.bloomberg.com/news/articles/2014-10-06/chile-top-renewables-market-on-sunny-desert-windy-shores

[75] *PV Grid Parity Monitor Residential Sector – 3rd Issue*. 2015, CREARA.

CHAPTER 14

CHINA

Many people cite China's emissions and energy consumption as a reason for inaction on climate and energy. The story goes something like: "What is the point of us doing anything when China is opening a new coal power station each week?" It is true that the country's coal consumption is the largest of any nation on earth; in fact they have used as much coal in their development as the rest of the world combined. It is also true that they are the largest overall users of energy, but there are always two sides to a story and this one has an interesting twist.

China is also the world's largest investor in renewable energy. In 2014 almost 29% of the world's total renewables investment, some US $89.5 billion, was spent by the country.[76]

As a result, China now has more wind power operating than the total capacity of the entire UK energy system, four times the total wind power installed in Denmark.[77] During 2014 it added 20.7GW taking its total up to 96GW of wind energy capacity. This makes it the world's largest wind energy market. Wind energy is now China's third largest power source behind coal and hydropower.[78]

China has also been the world's biggest solar market for two years running. China installed in excess of 10.5GW of solar in 2014. At the end of 2014, the total installed PV capacity connected to state grid was 28.05GW. Whilst this still lags behind the total installed capacity of Germany the speed of change is astounding. Electricity generated by PV showed rapid growth in 2014, increasing more than 200% over last year.[79] Solar generation capacity rose from 0.08-2% over the last four years. That is twice the share from nuclear power.[80] Six of the world's 10 largest solar panels makers, as well as five of the

[76] 'China Accounts for 29% of World's Total Renewable Energy Investment', www.theclimategroup.org/what-we-do/news-and-blogs/china-accounts-for-29-of-worlds-total-renewable-energy-investment

[77] 'China Rushes to Harness Wind While Government Still Pays', www.bloomberg.com/news/articles/2014-10-29/china-rushes-to-harness-wind-while-government-still-pays

[78] 'China's Wind Power Capacity Now Bigger Than UK's Total Electricity Supply', www.businessgreen.com/bg/news/2391764/chinas-wind-power-capacity-now-bigger-than-uks-total-energy-supply

[79] 'China: PV installed Capacity Grows to Almost 30GW in 2014', www.pv-magazine.com/news/details/beitrag/china--pv-installed-capacity-grows-to-almost-30-gw-in-2014_100018231/#ixzz3VLsi98Xc

[80] 'China Hunger for Clean Energy to Leave No Rooftop Behind', www.bloomberg.com/news/articles/2014-11-12/china-hunger-for-clean-energy-to-leave-no-rooftop-behind

top 10 wind turbine manufacturers, are Chinese companies.[81] Lots of the solar has been deployed in large-scale installations and now the focus is starting to shift into deploying on buildings.

Coal is still the dominant energy source in China, however. Coal generation made up 64% of China's electricity mix in 2013, and it is widely used in industry for steel production and other things. New coal-fired power stations are still being built, but the rate has slowed. More interestingly the amount the existing coal power stations are used has gone down to its lowest rate in decades. They are running at just 54% of their rated output. One explanation for this is the command and control nature of the Chinese economy. It has been built on central plans and investment decisions – so the power stations have just kept on being built when maybe they are not even needed. The low utilisation rates of these power stations would suggest that the policy makers have yet to catch up with a subtle shift in the market for coal-fired electricity.

Changes in China's coal output and consumption also point to this. For the first time in more than a decade, coal output fell by 2.5% and coal consumption fell by 2.9% respectively. China's biggest coal mine posted a 20% fall in 2014 net profit in 2014. In fact 70% of China's coal mines are losing money. They are basically producing too much coal as demand is slowing. Coal prices fell by 16% in 2014.[82]

These falls could not come sooner. The burden of pollution – 'airmaggedon' – across communities throughout China is exacting a heavy price on the health of the people and the environment, for the country's development. Many Chinese cities are suffering under thick blankets of toxic smog caused in part by their power stations. Understandably this is becoming a serious issue for people as they realise the health impacts of the situation. As one correspondent put it: "It's as if the 21-million-strong population of the Chinese capital is engaged in a mass city-wide rehearsal for life on an inhospitable planet. Only it's not a rehearsal: the poisonous atmosphere is already here."[83]

In a recent film on the subject called *Under the Dome*, investigative journalist Chai Jing went into details of the vast extent of the air pollution problem across China and the reasons why most environmental standards are not enforced. The situation for ordinary people is that they are living in a highly toxic environment, created by coal and oil pollution. Whilst people are very aware of the smog, most people are unaware of the health impacts. The 150 days plus of highly toxic air that occur each year in many major cities in China is very bad for human health. It is particularly bad for the most vulnerable – children and the elderly. The long-term effects on the health of

---

[81] 'Developing Countries Begin to Take Lead in Green Energy Growth', www.ft.com/cms/s/0/4832b922-5de8-11e4-897f-00144feabdc0.html

[82] 'China Shenhua Energy 2014 Profit Falls 19 Percent on Coal Price Slide', www.reuters.com/article/2015/03/20/china-shenhua-results-idUSL3N0WL25A20150320

[83] 'Inside Beijing's Airpocalypse – A City Made 'Almost Uninhabitable' by Pollution', www.theguardian.com/cities/2014/dec/16/beijing-airpocalypse-city-almost-uninhabitable-pollution-china

the hundreds of millions of people breathing this toxic air may take decades to become apparent.

The environmental legislation has not been enforced, so companies creating the pollution disregard it and don't even use the cleaner technology that in some cases is installed but sits idle. Chai Jing advocated individuals getting involved and reporting companies that were polluting in an effort to tackle the problem. The film was watched hundreds of millions of times in the week of its release, but then was effectively censored by the authorities.

People across the country have begun protesting the situation as awareness and tensions rise.[84] They are having some success in blocking some new developments and this is a major thing in a country where protest is seen as a threat to national security. It seems the authorities in some cases are starting to listen.

In Beijing, the city has shut down three of its four coal-fired power plants as part of its campaign to cut pollution. The final one is scheduled to close next year.[85]

One survey from 2013 of public attitudes on environment found that 80% believe that environmental protection should be a higher priority than economic development, and a similar number would be willing to protest to protect the environment in their area.[86] So the protests look set to continue. Perhaps they will be the key driver that pushes China onto a clean-tech path.

In March 2015, things took another step forward when President Xi Jinping, speaking at the annual session of the National People's Congress, pledged to punish violators of environmental laws with an 'iron hand'.[87]

This follows on from the pledge in December 2014 by President Xi Jinping to cap the nation's CO_2 emissions by 2030 or earlier if possible. This is a very encouraging development from a country that had previously only pledged to slow the rate of growth. As part of the same deal it has also promised to increase its use of energy from zero-emission sources to 20% by 2030.[88]

What this means for solar, for example, is in excess of a threefold increase in the technology's deployment to 100GW installed by 2020. This is similar to the entire world's installed capacity in 2013 and a huge jump.

So it seems likely that people power will play a significant role in the energy revolution in China. People who are no longer willing to breathe toxic air are

---

84 'How Will China Deal with Growing Anger Over Pollution?', www.aljazeera.com/indepth/ opinion/2014/07/china-pollution-protests-2014729105632310682.html

85 'Chinese Capital Shuts Third Coal-fired Plant in War on Smog', www.reuters.com/ article/2015/03/20/us-china-pollution-beijing-idUSKBN0MG1D120150320

86 'Kunming Pollution Protest is Tip of Rising Chinese Environmental Activism', www.theguardian. com/environment/chinas-choice/2013/may/16/kunming-pollution-protest-chinese-environmental-activism

87 'China's Pollution Assault Boosting Solar, Electric Vehicles', www.bloomberg.com/news/ articles/2015-04-08/china-s-pollution-assault-boosting-solar-electric-vehicles

88 'US and China Strike Deal on Carbon Cuts in Push for Global Climate Change Pact', www. theguardian.com/environment/2014/nov/12/china-and-us-make-carbon-pledge

starting to demand change and are being listened to. It seems that the leaders of China have decided to face up to the country's environmental woes and to go renewable. You can be certain that when they set their minds to it they will achieve their aim. It seems the first steps on this journey are well underway.

CHAPTER 15

INDIA

There are 1.2 billion people living in India and around a quarter of them lack access to electricity – over two thirds are using dung or wood fuel to cook. That's 300 million people with no access to electricity, more than the populations of the US and Canada combined. Over 800 million people are exposed daily to the hazards of indoor air pollution, with all the negative health impacts it has, just by cooking their food. Over the years many different governments have promised access to energy for all but none of them have yet been able to deliver on it.[89] It seems like renewable energy, and in particular solar, may just offer the chance to transform this situation for the better.

India was the fourth largest energy consumer in the world in 2011. Over 40% of their primary energy came from burning coal. Another 20% came from burning oil, most of which is imported. Like many other developing countries, lots of people are simply not connected to the electricity grid. Where the grid does reach, there are challenges with illegal hookups, bill fraud, or non-payment. To add to the challenges 15-30% of the power generated is stolen. To top it all off total power demand is on the increase in this growing nation.

As if the challenges of indoor air pollution were not enough, burning coal is inevitably having an impact on air quality. Recently New Delhi was ranked as the most polluted city on earth with air pollution levels up to 60 times higher than what is considered safe. According to the World Health Organisation (WHO), 13 of the world's dirtiest 20 cities are now in India.[90] A serious set of challenges that need to be overcome. The current Prime Minister, Narendra Modi, has said that his government wants to provide every home with access to power by 2019[91] – to bring 300 million people out of the darkness. That's a huge aspiration and he has a very short time to achieve it. Modi is aware of the falling cost of solar, calling it a 'game changer' as it starts to compete with conventional forms of electricity generation.[92]

89   'Modi to Use Solar to Bring Power to Every Home by 2019', www.bloomberg.com/news/articles/2014-05-19/modi-to-use-solar-to-bring-power-to-every-home-by-2019
90   'India's "Airpocalypse"', www.thediplomat.com/2015/03/indias-airpocalypse-learn-from-china/
91   'Government Aiming at 24x7 Power Supply: Narendra Modi', www.dnaindia.com/india/report-government-aiming-at-24x7-power-supply-narendra-modi-2011280
92   'Clean Energy Push: Drop in Solar Power Cost a Game Changer, says PM Narendra Modi', http://articles.economictimes.indiatimes.com/2015-02-16/news/59196954_1_solar-energy-renewable-energy-clean-energy-capacity

One thing that is different about Modi than all previous Indian leaders is that he has form on this front, as he has had real success on the energy agenda in his own state. The state of Gujarat where Modi was chief minister is now home to Asia's biggest solar park. During his time there, as well as driving this mega solar project, he also established a feed-in tariff scheme to put solar on city rooftops and canals. The Narmada 1MW pilot canal-top solar power project is the first of its kind. It generates 1.6 million units of clean energy every year whilst at the same time helping to prevent the evaporation of nine million litres of water from the canal. Gujarat now has an installed solar capacity of over 900MW, which represents over a third of the total solar power capacity installed in India.[93]

India had 249GW of installed electricity generation capacity connected to the national network in early 2014. Coal was providing 59% of the country's electricity needs in 2014. Renewable energy including hydropower was providing 29% of the nation's electricity.[94] As with all nations, electricity production is only part of the story, and plenty of fossil fuels are used in other areas of society, like industry.

In an effort to boost the renewable electricity percentage and start to transform access to energy in the country, the government has set an ambitious target of installing 170GW of wind, solar and biomass power projects by 2022.[95] It will be a massive jump if they reach this goal – but even getting half way there will probably be transformational. Some commentators speculate that this big target will be exceeded as the costs of renewable energy technologies continue to fall.

There may well be new coal plants built as well to meet the growing demand. But there is a battle of sorts around coal use in the country and that makes it a potentially tricky environment for investors. Much of the country's coal reserves lie under forests and on tribal lands, and communities in some affected regions are coming together to attempt to block access to their lands. Partly as a result of this, and some other factors like a struggling transportation system, coal imports have increased dramatically. So by 2012, coal imports into India had risen to 23%, from nothing in 1990. The extent to which new coal projects and new coal power stations will be developed almost certainly correlates to how fast renewable energy can be deployed. The faster renewables are added to the network the faster they will eat into the profit margins of the existing energy incumbents. Reduce the running time of the existing power stations and they become less profitable, share prices go down, and investment becomes harder to raise.

[93] 'Narendra Modi Plans To Bring Solar To 400 Million People, Electrify Rural India', www.cleantechnica.com/2014/05/26/narendra-modi-plans-bring-solar-400-million-people-electrify-rural-india/

[94] 'US Energy Information Administration – India', www.eia.gov/countries/cab.cfm

[95] 'Clean Energy Push: Drop in Solar Power Cost a Game Changer, says PM Narendra Modi', http://articles.economictimes.indiatimes.com/2015-02-16/news/59196954_1_solar-energy-renewable-energy-clean-energy-capacity

To reach the many people not connected to the electricity network, distributed renewable will be a key tool. The Indian government is seeing solar and other renewables as key in achieving this aim. In fact India is already leading in the area of mini-grids and off-grid applications. There are around 750 mini-grid systems in India, including 135 biomass rice husk gasification systems and 599 solar photovoltaic mini-grids, with a total capacity of 8.2MW and typically 10-400 customers each. This is a model that is set to grow dramatically.

The opportunities are huge in this rapidly developing country, and renewable energy is set to grow dramatically over the coming years. In a recent power auction the winning solar bids are providing power at rates that are cheaper than imported coal-fired power stations can achieve.[96] Surely this is the start of a massive change.

---

[96] 'Solar Cheaper in India than Imported Australian Coal', www.reneweconomy.com.au/2014/solar-cheaper-in-india-than-imported-australian-coal-60317

CHAPTER 16

JAPAN

In Japan they have perhaps had one of the most severe drivers to change their energy habits in the form of the catastrophic Great East Japan Earthquake and subsequent tsunami, which caused catastrophe across the country and in March 2011, resulted in the meltdown at the Fukushima Daiichi nuclear plant. This disturbing chain of events sparked an equally extraordinary response from a whole nation.

By February 2012, due to safety concerns as well as public opposition to nuclear power post-Fukushima, only three of the 54 nuclear reactors that produced some 25% of Japan's energy before the disaster were still active. This is a massive drop in power generation that could have brought an energy dependent economy such as Japan to its knees with disruptive rolling blackouts.[97]

What happened instead was quite remarkable and should be a source of inspiration to us all. The Japanese government launched the 'Setsuden' ('Energy Saving') movement, which called on the population to reduce their use of electricity by 15% during peak hours. This action was taken to prevent excess demand that could lead to a full system crash and ensuing blackouts, which would be very disruptive.[98]

The whole of the population has had to become engaged in energy saving, and this was achieved by top-down leadership as well as simple education. The general population had a real reason to change following the disaster – giving out live energy usage data in the form of a traffic light system on national television, as well as the internet. The state of play with the energy system was graphically represented for ease of understanding, with green indicating that the power usage is within 90% of the utilities' total capacity, yellow indicating that 90-95% of capacity is being used, and orange telling customers that they are using up to 97% of available energy. Red was reserved for moments when demand was close to peak supply.

Large-scale users of electricity were required to cut usage by 15%, with 100,000 energy reduction plans drawn up and enacted in the process. Wide

---

[97] 'Decrease of Japanese Power Consumption to Adapt to the Fading Nuclear Activity', www.leonardo-energy.org/decrease-japanese-power-consumption-adapt-fading-nuclear-activity

[98] 'Has Saving Energy Become the New Normal in Tokyo?', http://rendezvous.blogs.nytimes.com/2012/10/29/has-saving-energy-become-the-new-normal-in-tokyo/

ranges of measures were undertaken ranging from increased homeworking, altering factory shift times, casual clothes in the office in the summer to reduce cooling requirements, to the reduction of neon signage. Artists across the country produced a range of images to carry the messages widely.

The most amazing thing of all is that they achieved it. People engaged, and through the collective efforts of individuals, businesses and the utilities, the reduction targets were met within three months. What's more these were the hottest, most energy-intensive months of the year.

Japanese Environment Minister, Ryu Matsumoto, had said that the reduction would not be a temporary measure, but an event to change people's lifestyles. It appears he was right.

In various surveys that were compiled after the event, people's attitudes had shifted, not just to energy, but to many other things as well. People began to show more interest in having an energy-efficient home, even with a generator on it like PV. Some said, "I was surprised to discover many things that I can do without. I realised that I had been using too much energy."

The experience changed their relationship with food – they prepared simpler meals and wasted less. Profoundly, 80% of respondents in one survey answered that since the earthquake, "I came to realise the importance of nature", "I value a sense of togetherness with others", or, "I realise more than ever that people cannot survive alone".[99] With this new awareness, change becomes easier. Collective action becomes an obvious path.

To discover these important things in such an immediate crisis is perhaps expected. Most of us in the developed world forget that we are in fact in such a crisis, even if the waves of the climate tsunami have yet to break on our shores. We need somehow to remember this without being paralysed by the fear and enormity of it all. To take heart that meaningful change is not only possible, but can be an enriching experience, where we will discover important truths that have become obscured by our normal living. Let us all draw on the learning of the people of Japan to inform our next steps.

The Japanese government's next step was to embark on an ambitious renewable energy roll out. To make this happen a feed-in tariff was introduced to increase the uptake of the technology. Between July 2012 and October 2013, Japan added clean energy equal to the output of nearly four nuclear reactors. The feed-in tariff programme that was launched added 3,666MW during the year and permitted the construction of another 23,000MW of clean energy production. This is larger than the total installed capacity of clean energy technology that was deployed in the country before the programme began which totalled 20,600MW.[100]

[99] 'How Japanese Lifestyles and Awareness Changed after the March 11 Great East Japan Earthquake', www.japanfs.org/en/news/archives/news_id031476.html

[100] 'Japan Added 3,666MW of Clean Energy Since Incentive Program', www.bloomberg.com/news/2013-10-04/japan-added-3-666-mw-of-clean-energy-since-incentive-program.html

Much of this new construction was solar PV projects. Both domestic householders and businesses looked to the technology to provide for their energy needs.[101] Ironically in the shadow of the stricken nuclear plant, both big business and community have looked to solar to develop new energy sources. Toshiba announced that it will build Japan's largest solar project – a 100MW PV plant close to one of the towns evacuated in the disaster,[102] and a community group also established a Renewable Energy Village to demonstrate the concept of 'solar sharing' where they combine farming with the generation of energy by mounting solar above the farm land, whilst keeping it in production.

The process has inspired some new innovations like the Kagoshima Mega Solar Power Corporation that has built a 70MW PV system on reclaimed land. Commissioned in October 2013, this system will produce enough energy to power 22,000 homes.[103]

Another offshore energy project is stationed about 12 miles off the coast of Fukushima and is a floating wind turbine project, which was switched on in November 2013. Yuhei Sato, governor of Fukushima, said of the project: "Fukushima is making a stride toward the future step by step, floating offshore wind is a symbol of such a future."[104] Floating turbines have also been trialled in Norway and Portugal, but they may be very appropriate in Japan as it is surrounded by deep oceans where traditional turbines would be hard to install.

Nissan, who had launched the 'LEAF' electric car, also released a system to allow the electricity stored in cars' batteries to be supplied to your home by connecting the car to the house's electricity distribution system via the car's charging point.[105] This is one way to overcome grid intermittency issues, by using your car as the backup device for your house. This is an idea that has been much talked about but is finally becoming a reality due to these exceptional circumstances. Is this perhaps the precursor of the distributed storage system that we will need to build to smooth the output from our renewable generators?

In early 2014, the Mayor of the Fukushima province pledged to switch to 100% renewable energy by 2040 in a move that goes counter to national government stance, which is keen to see the nuclear power stations generating

[101] 'Japan Adds Nearly 4GW of PV Capacity', www.pv-magazine.com/news/details/beitrag/japan-adds-nearly-4-gw-of-pv-capacity-_100013954/#ixzz2qaBKxdi0

[102] 'Toshiba to Build Japan's Biggest Solar Plants in Fukushima', www.reuters.com/article/2012/06/20/us-toshiba-solar-idUSBRE85J0ET20120620

[103] 'Is Japan's Offshore Solar Power Plant the Future of Renewable Energy?', www.smithsonianmag.com/innovations/Is-Japans-Offshore-Solar-Power-Plant-the-Future-of-Renewable-Energy-180949453

[104] 'Fukushima Floating Offshore Wind Turbine Starts Generating', www.bloomberg.com/news/2013-11-11/fukushima-floating-offshore-wind-turbine-starts-generating-power.html

[105] 'Nissan Unveils New Power Supply System Through Nissan LEAF', www.nissan-global.com/EN/NEWS/2011/_STORY/110802-01-e.html

again – counter to popular feeling.[106] A survey of the population by the national broadcaster indicated that 80% of the population are opposed to restarting the nuclear power stations, however they voted for a pro-nuclear government.

Community-owned renewable energy is still in its early stages. Tetsunari Iida set up the first community energy project in Japan:

> The first project I worked with in 2001 soon after I came back from Sweden; I had read about community wind, like in Denmark. So I collaborated to set up a first community wind project near Hokkaido in the north end of Japan, it was rather successful. We created a community bond scheme under the Japanese financial system. It was a small but break-through project creating community wind as a financial system.

He has since gone on to become Executive Director of the Institute for Sustainable Energy Policies (ISEP), and in this role is seeking to seed the rise of community energy across Japan:

> We are trying to accelerate community power all over Japan, so far we have about 50 projects. Maybe next year and future years we will be doubling and doubling this. More and more local people are starting up distributed power or community power. So the local people are encouraged to work up renewable projects providing local ownership for the local community. Last week we coordinated setting up the Japan Community Power Alliance combining more than 50 community projects in Japan and it is increasing, almost every month more people are starting up community power.

One such company, Houtoku Energy, is a micro-electric power company in Odawara. It was established in 2012 and combines 24 local firms, including a 'kamaboko' boiled fish paste manufacturer and a taxi operator, who collectively financed the project with 34 million yen ($341,000). Masahiko Shizawa, the vice president of Houtoku Energy, said: "We hope to offer 'Odawara-made' energy for the good of the local economy."[107] Bring it on!

---

[106] 'Fukushima to Use 100% Renewable Energy by 2040', www.rtcc.org/2014/01/31/fukushima-to-use-100-renewable-energy-by-2040/

[107] www.houtoku-energy.com/project/

CHAPTER 17

NICARAGUA

If you think the renewable revolution is going to be a slow and long drawn out process, then it is worth looking at the story of Nicaragua. Famous for earlier revolutions, there is another sort of revolution happening in the country today and it is renewable.

In 2014 Nicaragua got over 52% of energy generation from renewable energy. It was a very different story in 2006 when 80% of their energy came from expensive imported fossil fuels. Rolling blackouts were common as the price of oil reduced their ability to provide power to the nation. Nicaragua is not blessed with a local source of fossil fuels, and much of their oil was imported from nearby Venezuela.

However Nicaragua is blessed with abundant natural energy resources: tropical sun, steady winds, lakes and mountains, and 19 volcanoes. The energy transformation that has occurred in under 10 years is pretty staggering, and makes their target to achieve 90% of their electricity supply from renewable energy by 2020 seem very possible. They will flip their energy dynamic entirely on its head in barely less than two decades.

Over 20% of their generation in 2014 was from wind power, 15% from geothermal, and 9% from hydropower. They can be pretty certain to meet their 2020 target as they will bring a new hydroelectric dam project online in 2019 – Tumarín – which will add 253MW of capacity into the mix.[108]

The country has over six million people and more than 70% of the population has access to electricity. The number of people with access to electricity has been increased in the last 10 years as well, with some rural communities being served by renewable micro-grids.

Realising they had a problem with energy supply back in 2005, the government of President Daniel Ortega opened the doors for investment in renewable energy. They did this by providing a stable policy framework and promoting the country as a place to invest both locally and worldwide. They gave renewable energy companies a tax break and allowed them to import equipment without duties. Ten years on the transformation has been surprising and investment has poured into the country.

One such example is Canadian company Ram Power, which invested

---

[108] 'Nicaragua Proyecta Para el año 2015 Contar con 53.99% de Participación de Energía Renovable', www.mem.gob.ni/index.php?s=1&idp=174&idt=2&id=801

$425 million into an existing geothermal plant set between the Telica and Rota volcanoes near the town of San Jacinto. They rebuilt the plant and it now provides 10% of the country's electricity needs. There is potential for many more like this.

Sugar and rum production is a big part of the country's economy and the producers have started turning the pulpy residue left from sugar cane extraction, bagasse, into a fuel to power biomass electricity plants.

Perhaps the most exciting thing about the story is that getting to 100% renewable is not the sole aim. It is estimated that renewables could produce eight times the energy requirements of the country. Nicaragua is well on the path away from energy insecurity to becoming an energy-independent powerhouse. It is likely that it will soon be exporting energy to the wider region and only with the power of sun, wind, biomass and earth. A great model for others to follow.

CHAPTER 18

COSTA RICA

For the first 75 days of 2015, Costa Rica got 100% of its electricity needs from renewable sources.[109] The country has been supplying over 90% of its electricity needs from renewable energy sources for quite a few years. Former Costa Rican president from 1994-1998, Jose Maria Figueres, was a strong advocate of renewable energy, and they have made significant progress.

The country has chosen not to develop its offshore oil reserves, but instead to focus on being the world's first carbon neutral nation. It is great to know there is a country out there that has such an aspiration. It has abundant renewable energy resources, with hydro playing a dominant role in the electricity generation mix. Over 75% of the generation comes from hydro dams, with geothermal, wind and solar all playing an increasing role as well. Their dependence on hydro dams has meant that in years of low rainfall they have had to rely on expensive diesel backup generation. Diversification of the energy supply by adding solar, wind and geothermal will reduce this challenge.

Costa Rica is known for its rich and beautiful natural landscapes, strict environmental protection laws, and conservation of forests and natural environments. This has at times been at odds with their energy ambitions, as large hydro dams can have a pretty negative impact on natural environments and there were protests to stop some developments. Some of their geothermal resource also lies in national parks and there are questions as to whether they can be developed without damaging the local environment. The country is still reliant on imported fossil fuels for transport fuel, but they have shown that it is possible to provide local renewable power as well as protect the local environment. They will get 100% of their electricity from renewable energy by 2016.

At the Lima climate summit in 2014, current President Luis Guillermo Solís restated their commitment to becoming a carbon neutral country by 2021. Their next challenge will be to electrify their transport system and the President announced plans for the creation of an electric railway system to be complemented with a new bus system that will only use biofuels.

The effects of going 100% renewable are plain to see for people in the

---

[109] 'Costa Rica Powered By 100% Renewable Energy For First 75 Days Of 2015', http://cleantechnica.com/2015/03/20/costa-rica-powered-by-100-renewable-energy-for-first-75-days-of-2015

country. Costa Rica will lower its electricity prices by about 12% from April 2015 because its demand has been met by almost entirely renewable energy this year.[110] Win-win.

[110] 'Costa Rica to Lower Electricity Rates Due to Renewables Rise', http://af.reuters.com/article/energyOilNews/idAFL2N0WL1XJ20150319

CHAPTER 19

SAUDI ARABIA

In 2014 I ended up in Saudi Arabia on a 'Trade Mission' headed up by the then UK Energy Minister, Greg Barker. A group of us from the UK were invited to travel to the country with the minister to explore the possibilities of doing business there. We met with members of the energy industry in the country as well as officials tasked with delivering on their renewable energy visions.

It seemed like an odd idea for one of the world's largest oil producers to be considering solar in a serious way. But as Bloomberg put it: "The world's largest crude oil exporter aims to have 41,000MW of solar capacity within two decades."[111] On top of this they were also looking to build 21,000MW of wind, geothermal and nuclear power. This 41,000MW of solar is more than Germany's entire installed solar PV capacity and all this is to be achieved by 2032.

During the course of the trip it became apparent that this was not 'greenwash' or some nice idea but that the ambition was there in response to a looming crisis. They have a double challenge in Saudi Arabia despite their huge reserve of oil: increasing demand for energy, and a government that is heavily dependent on income generated by the export of fossil fuels. As much as 80% of their government spending is dependent on oil revenues generated by oil export.[112]

According to K. A. Care, the Saudi government department tasked with delivering the vision: "Unless alternative energy and energy conservation measures are implemented, the overall demand for fossil fuels for power, industry, transportation and desalination is estimated to grow from 3.4 million barrels of oil equivalent per day in 2010 to 8.3 million barrels of oil equivalent per day in 2028."[113] Considering that in 2013 they produced around 9.6 million barrels per day you start to see the problem. At the current rate of increase in consumption the kingdom could be a net importer of oil within 25 years!

Their domestic consumption of oil is monstrous – 3.4 million barrels per

---

[111] 'Saudi Arabia Plans $109 Billion Boost for Solar Power', www.bloomberg.com/news/2012-05-10/saudi-arabia-plans-109-billion-boost-for-solar-power.html
[112] 'Burning Oil to Keep Cool – The Hidden Energy Crisis in Saudi Arabia', Glada Lahn and Paul Stevens, December 2011
[113] 'Energy Sustainability for Future Generations', www.kacare.gov.sa/en/?page_id=84

day in 2010 is growing each year as demand for energy increases, at around 7.5% annually.[114] In 2011, Saudi Arabia's electricity demand was 210TW hours, or about 7,420kWh per capita. They were using a similar amount of electricity to Mexico's total consumption but more than three times as high per person. Their domestic oil use is highly subsidised with prices in 2015 at $0.16 per litre of gasoline versus the global average of $1.12.

A significant amount of this oil and gas is burnt to produce electricity. Around 80% of that electricity is used in buildings, and around 70% of all the electrical demand is used to keep buildings cool.[115] Burning oil to keep cool – as a study into the challenges faced by Saudi Arabia is aptly titled[116] – is a pretty crazy reality.

Whether they will achieve their vision over the coming years is not clear at this point as progress on deployment has been slower than hoped. However one thing is certain: watching Saudi Arabia as they progress will be an interesting exercise particularly given that the stakes are so high.

---

[114] 'Saudi Arabia Wants to Become a Major Producer of Alternative Energy', www.oilandgas360.com/saudi-arabia-wants-become-major-producer-alternative-energy

[115] 'Summer Demand Taxes Saudi Power Sector, But Kingdom is Working on Solutions', www.iea.org/newsroomandevents/news/2014/august/summer-demand-taxes-saudi-power-sector-but-kingdom-is-working-on-solutions.html

[116] Lahn, Glada and Stevens, Paul (2011). *Burning Oil to Keep Cool – The Hidden Energy Crisis in Saudi Arabia*. Chatham House.

CHAPTER 20

SWEDEN

Energy use in Sweden is undergoing a quiet revolution, with the use of renewable energy already meeting over 48% of its total energy supply – not just the electrical supply.[117] Since 1990 it has managed to not only increase its GDP by 57% but also reduce its carbon emissions by 20%.[118] In 1967 Sweden became the first country to establish an environmental protection agency, and in 1991 it introduced a carbon tax.

Thanks in part to empowered local municipalities and councils, change has been led at a local level and has spread right across the country. Today a good exemplar of this is that all inner city buses in all major cities run on locally produced biogas. Sweden has been the pioneer of bioenergy, as well as a few other things besides. They get over 30% of their energy from biological sources, like trees and their by-products. With tree cover of 65% and a very active forestry industry this is perhaps not totally surprising. Hydroelectricity also plays an important role, providing over 14% of the energy needs of the country. Wind power is playing an increasing role, as well as heat pumps, which are well established in the country. District heating is now widespread, providing heating for more than 50% of the households and has played an important part in reducing the use of oil for heating. Of Sweden's 290 municipalities, 270 are using district heating. Stockholm itself has not just extensive district heating networks spread throughout the city, but a district cooling one as well. These heating and cooling networks have a range of inputs from seawater to heat pumps that enable waste heat to be utilised, and excess cooling or heating to be stored.

Around one third of all of the country's municipalities are 'eco-municipalities', a concept first established in the 1980s in Finland, which spread to Sweden as well. These eco-municipalities work to share best practice and learning on the journey to becoming more sustainable communities.

One such community is on the Island of Gotland in the Baltic Sea, 70km

---

[117] 'Generating Power for a Sustainable Future', https://sweden.se/society/energy-use-in-sweden/

[118] 'Sweden – GDP, Energy Consumption and Greenhouse Gas Emissions', www.norden.org/en/ nordic-council-of-ministers/ministers-for-co-operation-mr-sam/sustainable-development/indicators-for-sustainable-development-1/sustainable-use-of-the-earths-resources/decoupling-of-environmental-pressures-gross-energy-consumption-ressource-use-and-generation-of-non-mineral-waste-from-economic-growth/sweden-2013-gdp-energy-consumption-and-greenhouse-gas-emissions

from the mainland.[119] A popular destination for summer visitors, the island has a winter population of 57,000, which increases in summer to 200-300,000. I spoke to Bertil Klintbom, Director of International Strategies, Municipality of Gotland, about progress on this island.

"Gotland started quite early with Agenda 21 in the 1990s and with our own plan for the future. We have a vision that goes on to 2025 where we say 'we want to be carbon neutral'. That is quite a challenge to achieve so we have been working with quite a lot of different methods on the island."

They are already well on their way.

First of all, we have district heating that is completely free of fossil fuels. It is powered by woodchip, heat pumps and other sources such as biogas. It is in the main city of Visby but also in the bigger towns on the island. The city of Visby is a World Heritage Site with a medieval wall around the old city. At least half of the inner city is now connected to the district heating which is quite an achievement. We have smaller systems out in the rural areas where we have contracted farmers to produce district heating for a group of houses, a school and a home for elderly people. Local entrepreneurs have invested in district heating systems.

One other thing that we have been working with here when it comes to heating and cooling is using seawater. We have a new library and university situated here in Visby and most of the areas are heated by the Baltic and cooled by the Baltic. The energy consumption is really low using the seawater for heating and cooling.

The district has also worked hard to bring down their electrical consumption and increase the energy efficiency of their buildings.

We have quite a lot of wind power and on a fairly windy day there's around 40-45% wind electricity in the power grid here. We have a connection to mainland Sweden, but the cable is filled up so that is a problem at the moment. We are promised a larger cable in 2018 and then there will probably be a lot more turbines here after that. We have around 180 turbines today, some of them are owned by companies and others are owned by co-operatives, people who have bought them together. Now they are actually taking down three old turbines and putting up one bigger new one. People who are interested in energy and environment like the turbines but some of the tourists don't. But we do know that the seals like them. We have a lot of offshore turbines and people said that the seals and the fish would be harmed by the turbines that are standing out in the water but actually now there are more seals than before.

---

[119] www.gotland.se/eco

Another thing here is biogas, locally produced biogas. We think that biogas is something for the island to really work with and we have good resources. We have residues from farming and from the forest here that could be used for half the cars and we have quite a lot of cars on the island. I think there are around 35,000 cars and lorries. So a lot of them could use biogas for fuel. The politicians wanted us to start with biogas driven buses in the city and I ran a procurement for biogas buses. It was a local company who won the competition. So we suddenly had biogas buses but no biogas. Then we went out for a second procurement for biogas and guaranteed that the region would buy a certain amount so a local company here was formed, actually a farmer and some other companies who joined together to win the contract. In the beginning we used the biogas from the sewage plant here and the company built a purification unit to make fuel out of it.

Then we started the public bus transportation in the city and now the company here have gone on to take it a step further; they don't use the sewage any more. We use that for the district heating. They have built a new plant using farming residues. It is actually manure from pigs and carrots that can't be sold because there are a lot of carrots on the island and some of them are not good enough for sale, and they also grow corn, maize and other energy crops. So they produce biogas and some of it is purified for fuel and some of it is used by the big dairy that makes milk powder. They used a lot of fossil fuels there before and now they have switched to biogas.

In the future we hope to have more and more cars and lorries running on biogas. Now 60% of the fleet of trucks that are collecting garbage on the island use biogas. The region here is converting our own fleet cars. When a new car is being leased or bought it must be a biogas car. The fuelling stations that we use are the fuelling stations where the public can also fuel up their cars so the public are buying many bi-fuel cars that have a small petrol tank and a big tank for gas.

Not only have they addressed heating, electricity production and transport, but they have an enlightened attitude to waste. They have zero waste to landfill and no waste incineration and you pay per kilo for your waste disposal. The compostable material is collected and there is a plan for that too:

One thing we haven't succeeded with yet, but that is very close, is that we want to use the compost material for making biogas. If you have good compost materials you can actually create value from them. So you could then stop the collection tax; the costs for leaving the material doesn't have to go upwards all the time. You create something that has a good value. That is on its way.

Now plans are being drawn up for large-scale solar plants on the island to complement the wind generation, and they are still focused on the fossil free goal by 2020. It is refreshing to hear such a holistic attitude being applied to waste and energy generation: simple, pragmatic and inspiring. It seems to be widespread across Sweden.

Jämtland County is a rural and very wooded county in central Sweden with a population of 127,000. It gets 100% of its electricity from renewable sources.[120] In 2012, they produced 14TWh of renewable electricity in the region and only actually consumed 1.4TWh. Most of this production came from hydro projects.

In the 1970s, 80% of the heating was provided by oil. By 2012, 91% of all their heating needs came from renewables, and only 5% came from fossil fuels.[121] Using the abundant wood resource to power district heating and combined heat and power, they aim to get to 100% in the near future. In Jämtland's sole city, Östersund, a biomass-fuelled combined heat and power plant supplies 25,000 households with renewable heat and some electricity. In the villages throughout the county smaller scale plants using similar technology supply heat. Biofuel for transport is also used widely with energy being provided in the form of ethanol as well as biogas.

It just seems like people in Sweden have understood the opportunity provided by local generation, both for security of energy supply, employment and local job creation, economic advantage as well as environmental improvement. Not only have they understood it, but they have grasped it and implemented it across cities, rural regions and island communities. As the leader in renewable generation across Europe, they have quietly shown the rest of us what is possible if you get organised and get on with it.

---

[120] International Study of RE-Regions, http://reregions.blogspot.co.uk/2009/10/jamtland-conty-council-sweden.html

[121] www.energikontoret.z.se

CHAPTER 21

UNITED KINGDOM

In the UK, 20% of the electricity came from renewable sources in 2014 – around half of that from wind turbines. Solar PV deployment in the UK almost doubled in 2014 alone, getting up to 5GW of installed capacity,[122] with 650,000 homes with solar PV installed on them and a rising number of field-based solar energy systems. The UK is home to the largest offshore windfarm – the London Array – which sits in the Thames Estuary and provides power for nearly half a million homes.[123] Huge leaps in the deployment of renewable energy have occurred in the last few years to get to this point, but considering it is the windiest country in Europe it's not that impressive a record. The UK still resides at the bottom of the league tables for renewable deployment in Europe.

Perhaps this is in part because of the UK's heritage. The UK is famous for being the birthplace of the industrial revolution and the home of the coalmines that powered it. Coal was the engine that drove the development of the UK and at the start of the 20th century there were 1,000 coalmines in the country. Hard to believe today on this overcrowded island that they would all fit. Coal mining became a dominant force in the economy and in politics, and there were still 200 pits open when Margaret Thatcher decided to end the dominance of the industry in what led to the famous miners' strikes of the 1980s.[124] Today, even though nearly 40% of the electricity used in the UK comes from coal, only a handful of mines are still operating. Most of the coal used today is actually imported from around the world.

The rise of oil and gas extraction from the North Sea began in the 1960s and grew steadily until peak production was reached in 1999. In 2012, 67% of the UK's oil demand and 53% of the country's gas requirements were supplied locally, but this figure is reducing every year as dependence on imported oil and gas increases.[125] It's only a matter of time until the country is entirely

---

[122] 'Solar Power in the UK Almost Doubled in 2014', www.theguardian.com/environment/2015/jan/29/solar-power-in-the-uk-almost-doubled-in-2014

[123] 'World's Largest Offshore Windfarm Opens in Thames Estuary', www.theguardian.com/environment/2013/jul/04/offshore-windfarm-opens-thames-estuary

[124] 'Britain to Have Just one Remaining Coal Pit after UK Coal Announces Closures', www.telegraph.co.uk/finance/newsbysector/energy/10740600/Britain-to-have-just-one-remaining-coal-pit-after-UK-Coal-announces-closures.html

[125] 'North Sea Oil: Facts and Figures', www.bbc.co.uk/news/uk-scotland-scotland-politics-26326117

dependent on imported energy, unless we have a coal revival. I won't be volunteering to go down the mines.

We are blessed with a wealth of renewable energy resources in the UK, but perhaps due to the presence of local fossil fuel supplies the progress on developing them has been painfully slow. Perhaps it's just a case of old habits die hard? Not only are we the windiest country in Europe, we also have extensive coastlines and as much sun as Germany, which has embraced solar in such a big way. Yet the history of mills, both wind and water, seems to have faded into distant memory for most. Instead there is a strong anti-wind campaign whilst energy is imported into the UK from around the world.

The development of renewables has been slow. The first commercial wind farm in the UK was at Delabole in Cornwall, and it didn't become operational until 1991. Its development was driven by one family's vision to do something positive.[126] The first community wind scheme that went live in the UK was the Baywind Energy Co-operative. Keith Boxer started work on Baywind in the mid-1990s. It is an iconic development and was the only community energy project in the country for many years. Keith started the project after travelling and working in Sweden and meeting some of the community energy pioneers from the island of Gotland. With their backing and encouragement, he returned to the UK to search for a site and a project. Harlock Hill became that first site, eventually home to five turbines with an output of 2.5MW. Whilst the project was originally constructed using debt finance, it was refinanced with community investment. At first they just aimed to purchase one of the turbines for the community at a cost of £600,000 but in the end two ended up being community owned with a total investment of £1.2 million. He talks of the experience:

> For me personally, getting through the planning process was the challenging bit. You are dealing with a lot of issues: you're doing something that's good for people; it's good for the community; it's good for the environment and a lot of people are saying: 'What are you doing? Go back home to Sweden, you're only here for our subsidies. If it was a good idea, we'd already be doing it.' I was ambushed in meetings, for example. At one house I was expecting to speak to the people in the household about the project and there was a whole posse there of people who wanted to lynch me.
>
> I remember going to the planning committee meeting in South Lake District Council, sitting there as we watched them discuss the application. Every single committee member said something against the project when they voted for it; that was a really interesting experience. There was a lot of posturing but they decided that out of all the applications in that part of the world they were going to have to deal with, they knew they were going

---

[126] Delabole Windfarm, The First Commercial Windfarm in the United Kingdom, www.delabole.com/windfarm.html

to have to let one or two through. They decided ours was a good project and gave it approval because of the way we did it.

At that time there was of course elation, but I think there was just as much elation when the first application for somebody signing up to buy a share came back through the post. We started getting letters as people starting saying, 'I would like to buy 500 shares', 'I would like to buy 10,000 shares'. It was exciting that people believed in what we were doing. It hadn't been done before and they didn't know us from Adam. How did they know we were honest and wouldn't take their money and run off to the Seychelles with it? That we were actually going to make this thing happen was probably one of the very tangible rewarding points.

I was trying to do something difficult. You've got to know where you want to go and why. You've got to understand the purpose behind what you're doing and have an action plan for getting there. And if you have those things in place, what they say is that synchronicity will come to your assistance, coincidences will happen. I am not remarkable, I'm just an architect. I'm no different from anybody else. It's just that I decided to do something.

It doesn't matter how big the thing is that you do. It is the fact that you are doing something in the right direction, that gives you hope. We can't rely on people in government to solve it for us.

I want to share with you this image and the text from inside the prospectus that we wrote for the Baywind first share issue. I wrote that based on what I had seen happen in Sweden. It is the story of these people on this little island in the Baltic Sea who have decided they are not going to just complain about nuclear power, they are not just going out with the placards any more. They are actually going to make the alternative happen, because they realise that if they don't it won't happen. So this little group of people, three or four of them got together and this tiny island became a place of pioneering spirit for alternative and renewable energy. But in a Swedish sensible way, not a hippy, flaky way. I told the story of how the community movement on Gotland had started and how the idea of community ownership was what we were trying to translate into the UK. I've still got that prospectus. And when I read it I think, 'Gosh that story is just as valid today as it was 20 years ago.' It wasn't just me, there were many others involved that helped to make it happen. I was fortunate to be able to plug in to that knowledge and experience and passion.

Keith had to put in two years' hard work to get the project developed, which is actually quite quick for a project of the size. Others then took it on to the construction phase and beyond. Today Baywind has gone from strength to strength and now owns six turbines on two sites.[127] The project at Harlock Hill is being repowered with the turbines being replaced with larger machines. The Baywind story and the people who had been key to its success spun out into new ideas and ventures, one of which was Energy4All,[128] a company established to help communities with their own community energy projects. Energy4All has gone on to help many other communities establish projects across the UK.

Before Baywind were powering to success Adam Twine, a farmer from Oxfordshire, set off on a community energy mission that was going to take 15 years to come to fruition. Hearing the journey Adam went on to make his vision a reality and create a community-owned wind farm on part of his land, I am amazed by his determination and perseverance. His journey started in the early 1990s: "I was aware of the story of Delabole; Peter Edwards was a farmer who sold his cows to pay for the wind farm. I was also aware of stuff that had happened in Denmark. The idea of the wind farm being community owned seemed really nice but a bit pie in the sky. That was the initial thought so I put the planning application in in 1992."

Putting in a planning application on a wind farm in the UK is not a straight forward process. It is expensive and takes a lot of time. Adam ended up doing it three times! The first time he got permission the turbine manufacturer that was specified in the application went bust, and so he had to go back to the start. By the time of the third application, the issue had become politicised, as strong feelings around wind turbines started to surface in the local area:

> It was around the time of the first Gulf war which was obviously about oil and then we had that heat wave in France when a lot of people died. It crystallised from a local issue to becoming a local expression of a much bigger issue. It all became about 'yes' or 'no', you can't sit on the fence, where are you with climate change? Where are you with local solutions? I found that very exciting as well as hard work. Eventually we got the permission by one vote.

Following what can only be described as a 10-year saga, which is an incredible feat of persistence on behalf of all involved, it then became the moment to raise the funds to build the project. An exciting moment:

> We then went for the share offer. No one had ever done a raise for £4.5 million. Then the Financial Services Authority (FSA) changed the rules (surrounding co-ops) just before the launch and that delayed us another three or four months. We had to re-do it. So we launched at a really naff time – the end of November in the run-up to Christmas. We were already running months late so we did it anyway. It was really quiet at Christmas

---

[127] www.baywind.co.uk
[128] http://energy4all.co.uk

and in January. What was very exciting was in the last 10 days of February seeing the curve of money coming in just grow and grow. We had no idea whether it was going to work or not.

They got there. The elation of successfully raising the required funds didn't last long, as soon after writing to all the investors saying that they were selecting contractors and cracking on, the turbine supplier informed them it was unable to fulfil the order, due to a rush of orders from elsewhere in the world. They tried to place orders for different turbines with other manufacturers, but it got to the point where the only course of action was to start returning the money raised to investors, as it seemed the project wasn't going to go ahead after all.

"That was one of the darkest moments where we are a year down the road, about nine months from raising the cash. I'd probably been working on planning for 10 years."

Then the original turbine manufacturer came back to them as it could deliver the project after all – but they had to place an order right away. Unfortunately, they had already returned £1 million to investors.

> We thought we'd go to shareholders at an AGM and say, 'look, either we give you all your money back or we have four weeks to get the money in and go for it. Do you trust us to achieve this? Will you back us?'
>
> I remember starting the meeting; we had had the reports and the questions started. There were pretty negative questions not saying specifically, 'You've cocked up, give us our money back', but coming from that sort of place, quite challenging. Then a guy at the back put his hand up and said, 'Look, what we've got here is a really exciting project, there's nothing like this in the rest of the country. We can't let this go, these guys have done a really good job, we've got to go for it.' The room burst into applause and it changed the meeting immediately and it suddenly became, 'How are we going to do it?' It was one of those brilliant moments. We went back then to raise the money in four weeks. We needed a million quid in four weeks. I thought we weren't going to do it because why would people trust us and send money back again. We sent the emails out and letters and then waited. Extraordinarily we got the million quid in from members in the time needed and put the order in. It was a real people power good news story in the end.

The Westmill Wind Farm Co-operative was commissioned in 2008, and its five turbines produce enough energy to power 2,500 homes.[129] A huge achievement.

Not content with this, in 2010, with the advent of the feed-in tariff in

---

[129] www.westmill.coop/westmill_home.asp

the UK, Adam saw the opportunity to repeat the process and establish a solar farm co-operative on his land, Westmill Solar Co-operative.[130] Whilst similarly fraught in many ways, it was a much quicker process, taking just a couple of years. Adam worked with a commercial developer to get the project through planning, but as their planning permission came through the government changed the support programme for solar and that would mean the project was no longer viable. They then had a mad rush to get the project constructed, which wouldn't have been possible if they were to go out and raise funding through a public share offer.

Fortunately for Adam, the commercial developer was able to broker a deal with a funder to construct the project with an option to buy them out one year later, which would give time to raise the funds from the community. The project was constructed and like many others in the UK at the time it was a rush against the clock to get it connected whilst the support mechanism was still in place. There was another rush when it came to the co-operative buying it out from the commercial developers – various reasons meant the fundraising was delayed:

"We've got six weeks to write a prospectus, write the share offer, give it to the FSA and raise £4 million – we said £2.5 million minimum and £4 million is what we're aiming at." They actually raised £6 million in the period. They also did a couple of new things in the process: one of the board members managed to get a local authority pension scheme to provide the debt finance required for the project, and they set the investment scheme so that it paid back the capital during the life of the project as well as a return to shareholders. I asked Adam what effect the two schemes had on the viability of his farm:

It is marginally profitable to grow cereals there anyway. It's very thin land. Without subsidy I couldn't have done it, so I would have to be doing something else, so it's made me much more commercially resilient as a business.

I feel that farming is a generational thing; you are just a tenant or a guardian. You are entrusted with the care of that bit of land for a future generation of farmers. For the future we need to be changing what we are doing now. Because actually what you're doing now is an awful lot worse than when you started.

I think the important thing to me is not the hardware, it's about changing hearts and minds and people. We can do this differently. You probably know the story about grandfather talking to his grandson in some traditional society and the grandfather saying that, 'In my heart are two wolves, one's a good wolf and one's a bad wolf and they are fighting and I'm pulled in both directions'. The grandson says, 'Which one wins?' and the grandfather says, 'The one I feed.'

---

[130] www.westmillsolar.coop

We all get involved in social activities because we want to make a difference don't we? I was lucky I had one on my doorstep and I didn't actually have to traipse down to London all the time or over to Palestine to do something.

The early nineties seem to have been a busy period for renewable energy pioneers in the UK. Another pioneer was Dale Vince, who in 1995 founded a company called Ecotricity.[131] An ex 'new age traveller' who started the business from a caravan, he has gone on to build a green electricity company serving over 150,000 people across the UK. The company has built 19 windfarms across the UK as well as a number of solar parks. It has pushed the boundaries as an energy supplier, slowly building both its customer base and its generation portfolio. He talks about how he got here:

> I got into renewable energy off the back of a 10-year stint on the road, using small-scale renewables, living on a hill. I saw the first windfarm built in Cornwall in 1991 and I thought I should drop back in to society as I could make a bigger difference. That was the epiphany.
>
> So I set out to build a windmill and that led me to realise I needed to be an energy company to get a fair price for the power. That's how the journey began. I love what we do, and I do believe we are making a difference. It's just worth doing.
>
> Ecotricity employs nearly 600 people now, so we must be doing something right. We have affected policy and other companies not just in Britain, but around the world. Every year goes by we get bigger and stronger and more able to do things we believe in. I hope to bring about change by demonstrating by doing rather than preaching or trying to persuade others. Just to get on with it.
>
> The challenge to onshore wind energy at the moment in the UK is planning permission. Under this government there have been three changes which have made it much harder to develop a windfarm, and then they have given Pickles (minister responsible for planning) a new job to call in all the applications. So you follow all the rules and they stop you at the last hurdle.
>
> But public opinion is with us – the Department of Energy and Climate Change's own surveys show that. Wind energy isn't unpopular – that's just bullshit. All this from the party of business. It is fundamentally dishonest and what gives politics the bad name it has. I am the kind of person that likes to right a wrong. I feel injustice in the world very strongly. I am not one to back away from something that gets me into battles.
>
> The great thing renewable energy has to offer is a democratisation of the

---

[131] www.ecotricity.co.uk

energy sector, because people can make their own energy – it's a fantastic thing. I see us running the whole country on renewable energy, it is just a matter of when and what pain we go through before we get there. I see a smart grid that has at its heart a lot of storage capability to be able to flex and blend the different forms of renewable energy. A smart grid and smart devices in homes to enable intelligent demand management. These kind of solutions will transform the grid along with the decentralisation that renewable energy brings. With it comes the transformation of our economy, because we won't be shovelling billions of pounds every year into global fossil fuel markets – just to burn it. We will be making the energy here and creating the jobs here. Maybe we can get there in my lifetime. It just requires political will. Politics either gets in the way or it enables.

More and more people are concerned about the environment, and they want to do something about it. Our job is to produce affordable, simple solutions that people can adopt to help them change their lives.

The introduction of a feed-in tariff in the UK was a hard-fought process, with industry, NGOs and politicians working together to make it happen. Its creation was one of those key moments where communities and individuals were suddenly incentivised to get involved with energy and it was a major turning point for UK renewable energy.

To get support across all political parties it was built by a few very dedicated 'back bench' MPs who worked hard to make it happen. Alan Simpson, then a Labour MP, was at the heart of it:

> The bulk of the credit for the creation of the FIT has to go to the NGOs. However, I put down an amendment in support of feed-in tariff in the energy bill and the whips were going slightly berserk, as they didn't want the amendment passing. I had got a list of just over 50 Labour MPs who said they would support the amendment. At the same time all the major opposition parties would put their members on three-line whips to support the amendment.

This would mean the government would lose the vote if it tried to oppose it.

> I took the list in to show Ed Milliband, the new Secretary of State at the newly created Department of Energy and Climate Change. I said to him, 'look, this is a numbers game; with the commitments from our MPs and the opposition parties, you are going to lose the bill unless you support this amendment. The first outing for this new department and you as Secretary of State, and you are going to lose.'
>
> Milliband said he was facing huge opposition from his department, as well as others, like the Treasury. But he managed to find a way to get it

through with a reduced generation threshold. This allowed Ed Milliband to come to the dispatch box and genuinely steer through an amendment in his name that was groundbreaking. So everyone came out winning. I think that it was a moment probably as significant as the commitment to carbon targets. Feed-in tariffs were the mechanism that took us forward into clean generation. It was possibly the most important moment in my political career.

The advent of the feed-in tariff has meant that the use of small-scale renewable energy sources started to rise dramatically across the UK. Solar PV was particularly popular and went from around 10,000 installed systems in 2010 to 650,000 in about three years. This growth has not been without challenges as changes in politics meant that the feed-in tariff was reviewed and changed repeatedly – over 10 reviews and numerous changes in just three years. I was involved with other groups in challenging the government over the scale and process of the cuts to the scheme, both in the media and the courts. In the end, the industry that sprung up to deliver these projects has shrunk back to what it was before the feed-in tariff arrived, with many bankruptcies in the sector as a result. However for people or communities looking to adopt renewable energy, there has never been a better time.

More recently and after many years of pondering the government introduced the world's first renewable heat tariff – the renewable heat incentive. In response to the European 2020 renewable energy targets, the scheme is designed to drive the uptake of renewable heat generation solutions like biomass boilers, heat pumps and solar thermal systems. It has resulted in a big increase in the numbers of biomass boilers being installed across the country, particularly in commercial settings. There have been a number of wood fuel power community energy projects like the John Cleveland College Community Woodheat Co-operative, which heats a college by the same name.[132]

Communities across the country have begun to look more and more for solutions. With the rise of Transition Towns, groups formed to look for solutions to peak oil and climate change. People have started to look at what they can actually do in their area. Using drivers such as the feed-in tariff, many communities have started to become active energy generators in their local areas in the last few years. However the constant changes to the regulatory regime make this a challenging task. The UK energy market is dominated by six large companies, the same ones that operate across Europe. It is hard for communities to sell the electricity they might generate to individuals, so most communities focus on generation projects.

My hometown, Lewes, was the second Transition Town, and from it sprang our project Ovesco – which I talk about in some detail later in the book.

---

132   Green Fox Community Energy, http://greenfoxcommunityenergy.coop

We started looking for places to generate renewable energy locally, as well as giving away grants for domestic renewable generation schemes on behalf of the local council. In the end we built the first PV system in the UK that was funded using a community share offer. Like the stories above, we launched with a government induced time pressure in the form of changes to the feed-in tariff. We ended up having six weeks to raise the funds and build the project before changes to the tariffs came in and would have made the scheme unviable. Pretty hair-raising stuff for a group of volunteers.

Across the UK there are now many inspiring community energy examples, like Tors Hydro, the UK's first community-owned hydro scheme that renovated an old weir with an Archimedes screw water turbine and began generating in 2008.[133]

Repower Brixton was the first scheme in London, and was working in a neighbourhood that is very different from the middle class community I was working in. Repower London was the organisation that created the Brixton project, which installed a community PV scheme on an inner city block of flats. Whilst this doesn't sound particularly amazing, the way they went about it and the response of local people was. Project founder, Agamemnon Ottero, talks about the experience and the team:

> Repower is a flat structure. I am beyond proud of the fact that we are equally balanced in age, we have people from their 20s to their 70s, we've got black, white, brown, yellow. We speak seven languages. We've got gay, straight, lesbian. It's a real dichotomy of artists, scientists, aerospace engineers, all of them have one common goal which is to create wellbeing. Practical solutions to wellbeing. Every single one of them has hope. Hope that it can be done.

Working on the premise that community energy is about creating health and wellbeing in the community and not just a bit of renewable electricity, repowering has been working hard to reach hard to help communities.

> In my solar powered elevator up to my green roofed glass tower, I came up with some great ideas, but they were not the practical solutions needed on low income estates in Brixton or Hackney. That area in Brixton where we were working was one of the highest areas of deprivation in the country. We asked what they needed, actually going and knocking on doors. If your premise is how can I help you, you don't go anywhere. If your premise is how can we work together to create wellbeing for you and your family, for us all, then it begins to happen. That subtle difference is massive. When you come up with solutions together, that's when change happens.

---

133 Torrs Hydro, www.torrshydro.org/TorrsHydro/About-Us.html

In every project we have done there have been challenges. We had a triple homicide murder while we were there, 60-120 fuel poverty related deaths on the estates that we are working on, high levels of crime, the highest levels of unemployment. Heat or eat syndrome is common on these estates. We are not selling solar panels to these people, because that's not what they are asking for. What they are asking for is jobs, education and how to reduce their energy bills. So we run educational programmes six months before we put solar panels on the roof. First you go out and knock on every door. You tell people: 'We are going to run a programme offering paid training and mentoring for young people, for unemployed people, for construction workers. We will run fuel poverty initiatives, free installations of energy efficiency measures in their homes.' Then you tell them, 'There will be an opportunity for us all to own a power station on the estate. You will be an owner, you will have a say as to how it is run and you will be in control of the financial revenue from that power station. You will have control whatever the council does.' That is all on the first day when you knock on the door. They don't come the first four times you knock, and the fifth time they are still not interested, but three months later you break through. The real amazing thing when you get that co-operative up and running in year one and pay out your dividend, people come, whether they have invested or not saying: 'I really believe in this, I was wondering do you think we could do a green roof, or a garden, or another energy project?' We don't set up an energy project and walk away, this is a long-term project. The beauty of it is, this is built into the company from the start. You know that you have a 20-year revenue stream that can be used by this community.

There are nine stages to doing this thing and one of them is a renewable energy co-operative, putting solar panels on the roof. Each one of the stages is inclusive, and not only gives them hope about reducing their energy bills, it gives them hope that they have the right to change things. It gives them the ability to educate themselves, and create income.

The output is wellbeing, the goal is wellbeing, the engine that runs it is renewable energy. When we teach young people about the different company and co-operative structures, we teach them what they do. Then they choose what they want it to be.

At Repower London we are working as an aggregator for knowledge to help others. We are setting up individual co-ops that own and manage themselves. We got 57% of people on an estate with no money investing in the scheme which is great, but what about the other 43%? The whole community has to benefit. We may have to create our own legal structures for this. The government are playing with these legal structures, like a cat with a mouse, because they realise how totally destabilising this is. It is a fundamental shift from a bogus capitalist system to a participatory democratic inclusive system. It's truly innovative, in that people can be a part of their own wellbeing.

With the projects in Brixton, Agamemnon and the team around him were building the model, and repeated the process with three separate co-operatives in different estates. Now Repower London is helping communities across the capital to tread the same path, slowly the impact amplifies and magnifies.

In Plymouth, a community energy revolution is underway, and this one has the backing of the local council. In fact what makes it interesting is that the local council was instrumental in getting the scheme started. A visionary leadership team in the council set the objectives and empowered their staff team – who were equally keen to make it happen. The council did community consultation and feasibility work before creating the new company, Plymouth Energy Community (PEC). Giving it both staff time and financial support has meant that the new company has been able to make quick and substantive progress in areas where other organisations have failed.

Alistair Macpherson, CEO of Plymouth Energy Community, talks of the combination of people that came together to make this happen:

> We had a situation where the politicians and the officers were saying the same things and complementing each other. None of this would have happened without the political support. I have worked in local authorities for 11 years and I have never been able to deliver at the speed we have been able to over the last few years. The politicians had an idea and we have been able to fuel their ambition. They didn't know that much about community energy, but they wanted to see more renewables and they wanted to see fuel poverty solutions. We worked to provide them with both. The political support is being matched by officer ambition, which is not often the case. The leader of the council is keen, the finance lead understands the business case, and there are two other cabinet members who are saying this is the right thing to do out of a cabinet of eight. That's pretty unique.

The company started by focusing on a range of areas: getting people switching to cheaper energy suppliers, helping to alleviate fuel poverty and helping people to deal with fuel bill related debt. Their unique position of being independent, not for profit, but backed by the council has meant that they have achieved results that other organisations would never have achieved.

Alistair described a few of the areas of success:

> We are a local organisation that has more trust than the utilities, and we can market things in a more innovative way than the council can. We see much higher rates of take up on the offers that we market than commercial offers would achieve, but also higher than the council would get doing it on its own.

We have a grant scheme at the moment focused on energy efficiency and we are seeing a 10-15% response rate to our direct mailings, which is really high for stuff just landing on the doorstep.

We have been successful in getting some grant support in some areas, around fuel poverty and energy efficiency. Our work has helped to have £60,000 of household fuel debt written off for individuals in the last nine months, and we expect similar amounts to be achieved in the coming three months. That's fantastic. The funding for this work would not have been available to the council. That it's coming from a community energy organisation is really powerful.

It helps us join the dots between investors in the more long-term generation projects and those suffering because of the cost of living crisis. Part of the response we are getting we think is because people are seeing that it's not just about new tech and investment. Our ambition is to get the revenues coming in from generation assets to pay for the fuel poverty stuff in the long term.

Working with the community and the most vulnerable in the community has helped build trust with local people. Being backed by the council has opened doors to areas that may not ordinarily be easy to access. Local schools were one of these areas as PEC were able to work with the council to overcome barriers that prevented solar being installed on the properties previously, due to complexities in the ownership and lease set-up. PEC's first share issue raised money to build 21 solar PV systems on a range of schools across the city. After successful completion of the first project they are now raising money for their second set of installations.

The council has now invested in PEC to deliver savings for the community in the future. The support now allows the organisation to scale up much faster than it otherwise would. Some councils are saying they don't want to give their roofs to community energy organisations, that they want to do them themselves. I don't think they need to choose – do both as we have done here.

The stories from communities across the UK show the diversity of approaches people are taking to reinvent the energy system in their area. There are many ways to achieve this and there is a role for many different sectors and individuals in our society. Since these first steps from the pioneers 25 years ago, we are still taking baby steps, there is a growing movement. Now there is even a government strategy to promote community energy and various funding streams to support new pioneers.[134] All of these definitely help, but

---

134   Community Energy Strategy, www.gov.uk/government/publications/community-energy-strategy

the challenges around ever-shifting policy and regulatory environments often make the task of developing these projects onerous.

For me the common theme through all of these stories is that these projects happen because of the determination of the individuals involved and in spite of the government, not because of it. If a community scheme in the UK cracks the way to supply electricity directly to consumers, this could transform the market. Could the UK be the home of the next community energy revolution? It is going to need some very determined folks to make that happen. Volunteers? Read on.

CHAPTER 22

100% RENEWABLE

The Stone Age did not end for lack of stone, and the oil age will end long before the world runs out of oil.

Sheikh Zaki Yamani, former Saudi Arabian Oil Minister

One hundred percent renewable is already a reality in some areas, but are these just special cases with the 'right' conditions or is this something that can be achieved anywhere?

It is my opinion that it is not just possible, but it is also a realistic option everywhere. I believe it is technically and financially achievable, and it has a whole raft of benefits, for individuals, businesses, local communities as well as the environment. It will look different in different places, depending on what is abundant locally, but essentially it will be possible where you live and where I live.

I am not alone in these conclusions. There have been many studies and reports suggesting the very same thing.

Writing in the *Scientific American* in 2009 Professors Mark Z. Jacobson and Mark A. Delucchi lay out a clear path to 100% renewable by 2030, not just for the US but for the whole world.[135] Their conclusions were: "A large-scale wind, water and solar energy system can reliably supply the world's needs, significantly benefiting climate, air quality, water quality, ecology and energy security. As we have shown, the obstacles are primarily political, not technical."

In 2012 woolly tree huggers, Price Waterhouse Coopers, in their report '100% Renewable Electricity – A Roadmap to 2050 for Europe and North Africa',[136] tackle the same problem in a slightly different way but essentially come up with the same conclusions. Interestingly they conclude that not only is it possible, but that actually 'a power system based on low-cost renewable technologies is likely to be able to provide countries with opportunities for substantial economic growth and consumers with considerable – and growing – cost savings compared to a business as usual approach'. So not only is it cheaper in the long run, but it will be beneficial for us end users financially too.

---

[135] 'A Plan to Power 100 Percent of the Planet with Renewable', www.scientificamerican.com/article/a-path-to-sustainable-energy-by-2030

[136] '100% Renewable Electricity: A Roadmap to 2050 for Europe and North Africa', www.pwc.ch/user_content/editor/files/publ_energy/pwc_percent_renewable_electricity.pdf

What's not to like?

One of the most recent reports on the topic, written by consultancy Ecofys for WWF, also concludes that 100% renewable by 2050 is possible using technologies we have available now.

It lays out a scenario that requires a reduction in energy consumption:

> Global energy demand in 2050 is 15% lower than in 2005. This is in striking contrast to 'business-as-usual' projections that predict energy demand will at least double. This difference is not based on any reduction in activity – industrial output, domestic energy use, passenger travel and freight transport continue to grow, particularly in developing countries. Instead, reductions come from using energy as efficiently as possible.

It also forecasts an increase in the use of electricity: "Currently, electricity makes up less than one-fifth of our total final energy demand; by 2050, under the Ecofys scenario, it accounts for almost half. Cars and trains, for example, will become fully electrified, while other energy uses (such as fuel to heat buildings) will be minimized."

The report goes into detail on the opportunities and challenges likely to be encountered on the path to 100%, and yes we will need a fundamental change of tack from business as usual. However in that change there are so many opportunities for communities, individuals, businesses and governments to participate and benefit. This is a wholesale transformation that will provide lasting benefits.

Elon Musk, Product Architect and CEO of Tesla Motors, put it like this in an interview on the subject:

> By definition we must move to renewable energy. How can one argue against that? To argue for (not adopting renewable energy) is to say we will eventually run out of energy and die, or civilization will collapse. Obviously we must find ways to produce energy in a renewable manner. The question is what pace we should go at, and I think logically we should go as fast as we can. Since we know we have to get there eventually it's better to get there sooner, and prevent the environmental damage. To play devil's advocate, maybe the environmental damage won't be that bad, but why take the chance?[137]

As we will see from some of the pioneers' stories, islands are often the first to adopt these technologies and aim for, or even achieve, 100% renewable. This is often because so many are dependent on imported and dirty fossil fuels, and the transition is therefore more obvious. They have more to lose more rapidly

---

137 'ElonMusk – Thoughts on Transitioning to 100% Renewable Energy', www.youtube.com/watch?v=HiOLan8J0cE#t=57

in the face of climate change. They also have more to gain immediately in terms of energy security and economic benefits. So if we need proof that it is possible to go 100% renewable these islands have already shown that it is.

It is not only small islands that are showing what is possible. Iceland is a country with over 300,000 inhabitants and it is already very close to achieving the 100% renewable energy aspiration. They do have some special resources – hot rocks and fast flowing rivers – which they have used to achieve this. They must also have had the special human resources and leadership to do things differently, because they have set off on a path and succeeded spectacularly. Iceland generated 100% of their electricity from renewable sources in 2013. Amazingly over 70% of their electricity consumption was used by heavy industry – mainly aluminium smelting.[138] Over 90% of all of their heating is provided by renewable energy.[139] In total over 81% of all their energy requirements currently come from renewable resources. They still use oil for aviation and some other transport needs. So, OK, they are not 100% for all of their energy needs yet. However as they have made a clear choice to develop their local renewable resources I am sure it is only a matter of time until they apply the same thinking to their transport system as well. You may be tempted to think that they are in a unique situation because of their amazing volcanic and geothermal resource. But they are not alone; there are many other areas around the world that have similar resources that could be exploited.

The resources available vary in different regions and that makes the mix of energy resources different, but it doesn't make the goal of going 100% any less achievable.

There are quite a few other islands that have achieved similar, albeit on different scales, but fundamentally without Iceland's natural resources. El Hierro, the most western of the Canary Islands located off the West African coast, produces 100% of its electricity demand with renewable energy. Using wind to generate the bulk of their energy needs, they also pump water into a dormant volcano and store excess energy for times when the wind is less abundant.[140] The island nation of Tokelau in the South Pacific also meets its electricity needs from 100% renewable energy.[141] The island of Bozcaada off Turkey in the northeastern part of the Aegean Sea produces around 30 times the electricity needed by its inhabitants. Using wind turbines to produce energy, it exports its excess production to the mainland, as well as using some of it to create hydrogen to store energy for times when production is less.[142]

In fact it is not just islands that have shown it is possible, there are also 100%

[138]   Energy Statistics 2013, http://os.is/gogn/os-onnur-rit/orkutolur_2013-enska.pdf
[139]   Direct Use of Geothermal Resources, www.nea.is/geothermal/direct-utilization/nr/91
[140]   The Wind-Hydro-Pumped Station of El Hierro, www.goronadelviento.es/index. php?accion=articulo&IdArticulo=121&IdSeccion=104
[141]   'Tokelau Aims for 100% Renewable Energy in 2012', www.undp.org/content/undp/en/home/presscenter/articles/2012/05/07/tokelau-aims-for-100-renewable-energy-in-2012
[142]   Island of Bozcaada, Turkey, https://go100re.net/properties/island-of-bozcaada-turkey/

renewable regions. The region Mureck in Austria already creates 160% of its energy demand for heat, electricity and fuel from renewable energy.[143] Using biodiesel, biogas, biomass district heating and solar PV they have managed to become a net energy exporter. Local people are actively involved, as some of the systems are owned by co-operatives.[144] Extremadura, in southwestern Spain, supplied the electricity demand for its over one million people entirely by renewable sources for the first time in 2010. It was a bumper year for wind, sun and rain which meant that they generated more energy than was expected, but have gone on to add more renewable generation since then despite the challenges faced by renewable energy in Spain.[145] A small Czech village called Kněžice has turned its waste issues into biogas and also built itself a biomass district heating system. In the process it has effectively become the first energy self-sufficient village in the country. The mayor of the village, Mr Milan Kazda, said of the endeavour: "I would definitely recommend this system to other municipalities, but I would encourage them to come over to our, or a similar facility, and learn from the mistakes that we have made through ignorance of the system."

Inevitably there will be challenges and we will get it wrong on the path to going 100% renewable. That is no reason not to believe it is possible or to start out on the path. Take heart that many already have started out and some have already got there. Draw strength from the many other communities out there who have blazed a trail for the rest of us. Go visit them, listen and learn; their hard won lessons will save us all many hours of hard work.

SHOW ME THE MONEY

Where money is flowing is often a good barometer of whether things are working, and money is certainly flowing into renewable energy. According to *Bloomberg New Energy Finance*, in 2014 total clean energy investment jumped 16% to $310 billion. This investment "represented a more than five-fold increase on the $60.2 billion achieved a decade earlier."[146] That's a pretty huge jump in 10 years – the money is flowing into this revolution with a growing pace.

Notably in recent years there have been some pretty high profile invest-ments. The 'Sage of Omaha' Warren Buffet, currently the world's third wealthiest man and one of the planet's most successful self-made men and

[143] Appreciation Municipality of Mureck, www.eurosolar.de/en/index.php/appreciation-municipality-of-mureck

[144] The Energy Cycle System Of Mureck, www.100-res-communities.eu/bul/communities/best-practices/the-energy-cycle-system-of-mureck

[145] 100% Renewable Electricity Region, www.go100percent.org/cms/index.php?id=69&no_cache=1&tx_ttnews%5Btt_news%5D=119&cHash=c4f6018fc11f213a07feea183601df78

[146] 'Clean Tech Investment Surges Back in 2014', www.theguardian.com/environment/2015/jan/09/solar-power-led-clean-energy-investment-surge-in-2014

investors, has so far invested $15 billion in renewable energy. When asked about this at a conference in Las Vegas in 2014 his response was clear: "There's another $15 billion ready to go, as far as I'm concerned."[147] Warren Buffet's various investments in wind and solar are a huge stamp of credibility for the sector, seen by people all around the world. His confidence to invest gives others confidence to do the same.

Today companies are often larger in turnover and impact than countries. Many of the companies that we use on a daily basis are embracing this revolution too and investing heavily to make it happen for them.

"At Google, we're striving to power our company with 100% renewable energy. We're currently using renewable energy to power 35% of our operations, and we continue to look for ways to increase our use of clean energy."[148]

"To date, we've made agreements to fund over $1.5 billion in clean energy wind and solar projects."[149] Their investments have been made on three continents with a capacity of more than 2.5GW. They currently offset the rest of their emissions to make their climate impact zero.

Apple scored pretty low on renewable energy in a Greenpeace report in 2011.[150] But by 2013 they had transformed their approach:

> Our goal is to power all Apple corporate offices, retail stores and data centers entirely with energy from renewable sources. We're designing new buildings and updating existing ones to use as little electricity as possible. We're investing in our own Apple onsite energy production as well as establishing relationships with third-party energy suppliers to source renewable energy. As of 2013, we've already converted 73% of the energy for all our facilities – 86% for our corporate campuses and 100% for our data centers. And so far in 2014, we're powering more than 140 US retail stores with renewable energy.[151]

Not bad going in two years!

IKEA is another company that has demonstrated it is serious about transforming its business to run on renewable energy, as it has done loads already. By late 2014, IKEA had installed 550,000 solar panels across its stores and had made investments in wind projects in eight countries – a total of 206 turbines. In 2013, IKEA produced the equivalent to 37% of the company's

---

[147] 'Warren Buffett Could Double Renewable Energy Investment to $30bn', http:// blueandgreentomorrow.com/2014/06/11/warren-buffett-could-double-renewable-energy-investment-to-30bn/

[148] Google Green – Renewable energy, www.google.co.uk/green/energy/

[149] Google Green – Investing in a Cleanenergy Future, www.google.co.uk/green/energy/investments/

[150] 'Greenpeace Praises Apple's Increased Commitment to Renewable Energy', www. renewableenergyworld.com/rea/news/article/2014/04/greenpeace-praises-apples-increased-commitment-to-renewable-energy

[151] Environmental Responsibility, www.apple.com/uk/environment

total energy (not just electricity) needs from renewables, around 1,425GWh. In late 2014 it was just completing the purchase of the Hoopeston wind facility in the US, which is expected to generate the equivalent of 165% of the electricity consumed by IKEA US. Not bad considering there are 38 stores, five distribution centres, two service centres and one factory. IKEA has invested heavily in energy efficiency as well.[152]

As reported in their group sustainability strategy, "By the end of FY15, we will produce renewable energy equivalent to at least 70% of our energy consumption and by the end of FY20 we, on a Group level, will produce as much renewable energy as we consume."[153] Steve Howard, Chief Sustainability Officer of the IKEA Group explained the rationale: "Investing in renewable power makes complete business sense. It aligns with our corporate expectations on financial returns and our values. We plan to invest approximately €1.5 billion in new renewable energy projects to meet 100% by 2020 goal – and RE100 is a great way to tell our story."[154]

As of November 2013, UK supermarket chain J Sainsbury's had installed its 107,396th solar PV panel creating the largest multi-roof solar array in Europe with an installed capacity of over 23MW.[155] They are also in the middle of a programme to roll out geothermal heat systems on their stores – with an expected 100 stores completed by 2016. They have installed 42 biomass boilers since 2008, and they send no food waste to landfill. In achieving this last aim they have become the UK's biggest user of anaerobic digestion.[156] This is a good start for a big bad supermarket.

There are of course corporates whose green speak doesn't seem to add up to real action to develop renewables and address their carbon footprint. Walmart is one in that camp that was criticised in a recent report by the Institute for Self Reliance: "Despite pledging nearly a decade ago to shift to 100% renewable energy, Walmart today derives only 3% of its US electricity from its renewable energy projects."[157]

Importantly there are many others that do take it seriously and are taking action. They are not doing it out of some altruistic drive, of course that is

[152] 'IKEA Continues Trend Of Supplying Its Own Renewable Energy', www.forbes.com/sites/peterdetwiler/2014/04/10/ikea-purchases-98-mw-us-wind-farm-continues-trend-of-supplying-its-own-renewable-energy

[153] 'People & Planet Positive: IKEA Group Sustainability Strategy for 2020', www.ikea.com/ms/en_GB/pdf/reports-downloads/peopleandplanetpositive.pdf

[154] 'Going 100% Renewable is a Smart Business Decision: IKEA, Swiss Re, Mars, BT and Other Leading Companies Pledge to Transition', www.theclimategroup.org/what-we-do/news-and-blogs/going-100-renewable-is-a-smart-business-decision-ikea-swiss-re-mars-bt-and-other-leading-companies-pledge-to-transition

[155] 'Our 20 Commitments to Help us all Live Well For Less – Our progress so far', November 2013, www.j-sainsbury.co.uk/media/1790641/20x20_brochure_2013.pdf

[156] 'Sainsbury's and Partners Roll out Renewable Energy to Supermarkets', www.j-sainsbury.co.uk/media/latest-stories/2012/20120719-sainsburys-and-partners-roll-out-renewable-energy-to-supermarkets

[157] 'New Report: Walmart's Dirty Energy Secret', http://ilsr.org/walmarts-dirty-energy-secret/

certainly part of the motivation, however what really drives this action is money. As IKEA's Chief Sustainability Officer put it: "Investing in renewable power makes complete business sense." Invest in renewables and hedge your energy costs, invest in energy efficiency and increase your profits. Simple equations that inspire action when money and return on investment are a key deciding factor. The fact that so many large corporates are taking these actions just proves a point – it makes sense financially, for the future of their business, for their public image and for the planet. What's not to like?

So the money is flowing into clean green local energy from many sectors, but what about your money? Your savings, your pensions, your investments. Where is your money and what business is it funding? Coal, oil and gas? Or is it in the hands of the traders and the markets? Shouldn't you know where your money is and what it is doing? The local renewable revolution gives you that option – you too can own a piece of this new system. Stable returns from systems and businesses doing useful work in your community, and creating positive change at the same time. It is the moment to move your money and turn it into a force for good, whilst helping to insulate your community from the price volatility of fossil fuels, and the mounting problems faced by climate change.

CHAPTER 23

BATTLE LINES

The choice is ours, and though we might prefer it otherwise we must choose in this crucial moment of human history.

Rev. Martin Luther King

Behind many of these stories of positive change lies struggle. None of the stories you have read have happened without significant effort and determination on the part of those involved. Challenging the status quo takes a lot of energy, determination, resilience and staying power.

The status quo is represented in the corporations that run our energy industry. Fundamentally they have one motive: profit. Whatever their adverts and corporate social responsibility pieces say about the 'triple-bottom line' and 'sustainability', the purpose of the corporate structure is making a return to shareholders and ultimately this trumps ethical considerations in all but the most responsible of businesses. Ostensibly there is nothing wrong with making profit, in fact it is essential for a viable business. However the profit motive in isolation is often at odds with many of the things we all need: affordable energy and a healthy planet for our children. That is where the problem lies; the profit only motive is preventing the progress and transformation we need at this point. Of course after 20 years plus of the 'corporate social responsibility movement', there are enlightened corporations out there grappling with these issues, but equally there are others that are simply focused on extracting the maximum returns at whatever the cost.

Germany has been a pioneer in challenging the energy industry status quo. Talking to Hans-Joseph Fell, former Green Party Member of Parliament who wrote the now famous EEG law which created the 'Feed-in Tariff' in Germany, I was keen to understand how Germany had achieved so much and what forces they were battling to make it happen.

The opposition to energy efficiency and renewable energy has a decades long history, where it is opposed every time. This is a worldwide thing. In Germany big energy had politics in their pocket until 1999. In 1999 we had an election where for the first time the Green and Social Democratic party were in coalition.

The lobby works worldwide with the same principles. Billions of euros for propaganda and direct lobbying. Paying for parties, paying for the people and when these allowed figures are not enough, then comes corruption. Worldwide corruption plays the biggest role to support and protect the business of nuclear, coal, oil and gas.

They have a lot of money. You should know that they do not bring business for the population. For example Exxon, the biggest mineral company in the world, has only 40,000 employees. It is nearly nothing. When they have so few employees, you can imagine how much money some people can earn within it and they can influence public opinion with this money completely.

According to Oxfam: "In 2013, fossil fuel industries spent an estimated $213 million lobbying US and EU decision makers – well in excess of half a million dollars every day and totalling $4 million a week. In the US alone, the estimated yearly bill for lobbying activities by fossil fuel interests amounts to $160 million."[158] When you consider the size of the subsidy collected by these industries each year, this lobbying investment pays dividends. This is of course the funds that can be tracked.

When it came to creating the EEG, the pioneering renewable energy law in Germany, Hans Josef-Fell had to fight hard with the opposition within his own coalition government:

> The minister did not work on the Renewable Energy Act. It came out from Parliament, it was proposed from Parliament and it was decided in the Parliament. That it was not a government proposal is very important because in the Social Democratic Party ministers were very opposed to it. In our own government we had opposition. The head of the opposition was the Economic Minister, Werner Müller. He came from the coal industry and afterwards he went back to the coal industry. It was behind the curtain, outside of the public view. It was a very, very hard fight against him and against some others of the Social Democratic Party. There is one man who carried a lot of other Social Democratic parliamentarians around him, Hermann Scheer. He died in 2010, but together we were the pressure, me in the Greens and him in the Social Democratic Party.

Battles like this are happening in many of the elected bodies around the world over energy and climate issues. The German experience was repeated in the UK towards the end of the last Labour government with the ultimately successful 2008/9 campaign for a renewable energy feed-in tariff. The renewable energy industry and its allies in the wider green movement faced an enormous battle to overcome entrenched Whitehall thinking and the energy industry

---

[158]   'Food, Fossil Fuels and Filthy Finance – 191 Oxfam briefing paper, 17th October 2014', www.oxfam. org/sites/www.oxfam.org/files/file_attachments/bp191-fossil-fuels-finance-climate-change-171014-en.pdf

incumbency, which lobbied hard to frustrate the feed-in tariff campaign and to maintain the status quo. As in Germany, a fundamental key to the success of that campaign was the role of many dozens of 'ordinary' Labour MPs led by campaigners such as Alan Simpson MP and Colin Challen MP, absolutely willing to challenge energy policy orthodoxy within their own party and to do so very publicly, despite obvious irritation from their own party leadership.

The battle raging over the Keystone XL pipeline, planned to carry the tar sands south to the US from Canada, is a case in point. A bill has passed through the US Senate that allows it to be constructed, and all that stands in its way is a veto from President Obama.[159]

There are also many examples of the collusion between government and oil, gas and utility interests. In the UK in early 2015, a letter was leaked that had been sent by the UK Chancellor, George Osborne, to a high-level cabinet committee on economic affairs. It was a detailed list of how the government needed to intervene in many areas to support the growth of the fledgling fracking industry in the UK. The chancellor demanded 'rapid progress' on 'reducing risks and delays to drilling' from other ministers.[160] The government is also funding an 'information' campaign for the fracking industry. Hardly a balanced approach, with government figures at the highest level pushing the agenda of a controversial industry and specific companies within it. Can you imagine the outraged headlines in the right wing press, if the government had decided to fund a pro-solar or wind information campaign at the taxpayers' expense?

There were repeated reports over the last few years of people on the energy industry payroll working inside the UK Department of Energy and Climate Change, working on the laws and regulations governing the functioning of the energy market – in fact developing new regulation in the area.[161] Not surprising then perhaps when the new law designed to transform the industry arrives in 2015, its first act is to deliver a multi-billion cash windfall for existing nuclear and coal-fired power stations.[162]

Conservative MP Laura Sandys, who was Parliamentary Private Secretary to the Minister of State for the Department for Energy and Climate Change, told me her thoughts on the utilities and their future:

> I think it is a busted business model, so they are trying to do anything they can to maintain it. It is not really bust for another 10 years and they are trying to extend that to 20 years. They are not trying to kill [renewables]

[159] 'Senate Passes Keystone XL Bill', http://columbiadailyherald.com/news/local-news/senate-passes-keystone-xl-bill

[160] www.theguardian.com/environment/interactive/2015/jan/26/george-osborne-fracking-letter

[161] 'Gas Industry Employee Seconded to Draft UK's Energy Policy', www.theguardian.com/environment/2013/nov/10/gas-industry-Employee-energy-policy

[162] 'EDF Stands to Rake in £3.3bn Windfall from Existing Power Plants', www.theguardian.com/business/2014/dec/14/edf-3bn-windfall-existing-power-plants

but they are trying to put on every brake that you can possibly imagine, because they haven't got a solution to their business model yet.

This seems borne out in so many examples around the world, where fledgling renewable energy industry is suddenly stopped in its tracks.

In the UK, we have had very damaging reviews of the schemes promoting solar energy which I ended up being very involved campaigning against.[163] These knee jerk changes to the support schemes that supported solar ended up with many job losses and bankruptcies in the sector. During the feed-in tariff battles of the last decade, it is a fact that companies such as SSE lobbied against the feed-in tariff, then argued that it should be restricted to 'micro-generation' only. In the wind sector, the Conservative-led coalition government has a strong anti-wind narrative despite regular government polling showing overwhelming public support for the technology, and Eric Pickles MP intervened in the planning process to reject wind farms that would otherwise have been approved.[164] All of this adds up to massive amounts of wasted investment and heartache for all involved – I can testify to that personally.

In Australia the headline says it all: 'Renewable energy target thrown into confusion as negotiations collapse',[165] as the government looked to reduce the country's renewable energy targets and the industry is beset with uncertainty. This is in stark contrast with the PM Tony Abbotts' strong backing for the world's dirtiest fuel, coal. "Coal is good for humanity, coal is good for prosperity, coal is an essential part of our economic future, here in Australia, and right around the world."[166] Oh dear.

These few examples are of the battles and actions that we can see. There are many others playing out behind the scenes that never reach the public consciousness. In fact the thing that often reaches public consciousness is that renewable energy is not good for us. According to one study more than half of the coverage of renewable energy solutions in the mainstream British press is negative.[167] An article by Amory Lovins sums it up nicely: "Mainstream media around the world have a tendency to publish misinformed or, worse, systematically and falsely negative stories about renewable energy."[168]

---

[163] 'Solar Subsidy Cuts Spark Job Fears', www.independent.co.uk/environment/green-living/solar-subsidy-cuts-spark-job-fears-6699359.html

[164] 'Eric Pickles Clamps Down on Onshore Wind Farm Applications', www.ft.com/cms/s/0/66eb2b74-e804-11e3-9af8-00144feabdc0.html#axzz3RTvYJC8r

[165] www.theguardian.com/australia-news/2014/nov/11/renewable-energy-target-in-confusion-as-negotiations-collapse

[166] 'Coal is the Future, Insists Tony Abbott as UN Calls for Action on Climate Change', www.theguardian.com/environment/2014/nov/04/coal-is-the-future-insists-tony-abbott-as-un-calls-for-action-on-climate-change

[167] 'How UK Newspaper Coverage is Skewed Against Renewables', www.theguardian.com/environment/blog/2011/aug/05/uk-newspapers-renewables

[168] 'Debunking the Renewables "Disinformation Campaign"', http://blog.rmi.org/blog_2013_07_31_debunking_renewables_disinformation_campaign

Jeremy Leggett, founder of Solarcentury and Solaraid, describes this struggle in stark terms:

It feels very much like a civil war to me, a long running civil war. There are two warring belief systems in conflict. The two belief systems are variably distributed in people across companies and governments.

For a quarter of a century the light side has been losing to the dark side and I have been on the losing side of many battles. Including within solar.

But it does feel like we are in danger of winning. We are beginning to win. So we have passed El Alamein. Like Churchill famously said we have reached 'perhaps the end of the beginning'. We can win this thing.

Now this is not the end. It is not even the beginning of the end. But it is, perhaps, the end of the beginning.

Winston Churchill

Jeremy continues:

The sad thing is that the battlefield is littered with corpses, and I fail to be convinced that this is all the so-called natural process of creative destruction in modern capitalism. This is really dysfunctional stuff. Perfectly good companies have gone bankrupt because of malign action in the energy incumbency.

Ultimately what we learn from all this is what Germany has learnt, that the future will lie substantively in people power. People power in the energy sector, people power in the financing of the energy sector, and indeed financing full stop. That is how I see the road to renaissance. That is what gets me out of bed in the morning. I know we have a much better chance now than at any time in the quarter century I have been working on it.

The politicians in Germany and other areas around the world rely on popular support to push change from the status quo. Without our backing, without a movement for change, without our leadership even, they lack the mandate to make the fundamental changes we will need to succeed on this mission.

In Germany a broad social movement provided that social licence. Hans Josef-Fell describes the movement that made Germany's progress possible:

Support comes from the wider population, anti-nuclear, anti-pollution, anti-global warming. This movement wants solutions, it does not want to be against things. In the 1970s and 80s people formulated a framework around renewable energy. And then in 2000, when the FIT law came into force, all these groups had political regulations so that they could make a profitable business with photovoltaic, wind power and others, and then it exploded. So it came together, the political regulations and the activity of the population.

The movement created a lot of new business models that we had not seen before. I mentioned I created the first co-operatives for solar, but today in Germany we have 900 of them. People learnt how to make businesses from renewables and not leave it to the big companies. The motivation for this was not only climate protection and nuclear phase-out. It was also to make new businesses, jobs and create taxes for our villages. The motivation was very great.

So in these times of choice, what will your choice be? Sit by and let it unfold? Or get stuck in and join the movement globally that is building solutions in our neighbourhood?

Our actions may determine if we become a casualty in the war for a habitable planet for generations to come.

Leonardo DiCaprio

PART III

Your Guide to Making It Happen

CHAPTER 24

THE JOURNEY

I believe that a billion acts of courage are critically needed now for us to create a better future.

Kumi Naidoo

Now you know what is possible, it is time to get stuck in.

I am convinced we all need to get involved with building the new energy system that our communities will need in the future. That could be as doers, leaders, investors, supporters, ambassadors or promoters. Getting communities involved in local energy generation is the only way to redesign our energy systems at the speed that they need transforming. This is not just in response to a threat, but it is a huge opportunity for individuals and communities. The shift to renewable energy is happening, and there is an opportunity for you and your community to take part in this great change, to benefit from it. This could make the difference between energy poverty and access to affordable energy.

In some cases politicians have had the vision to set the right conditions for these schemes and initiatives to flourish. In many others, it is communities and individuals that lead the change, and they have to fight through the bureaucracy to achieve their aims. Personally, having seen the power and impact of large monopoly energy businesses on the political process, I think it is essential that we, the ordinary people, get involved and lead the change at local level. This in turn will show the politicians what is possible and what we want. It will also take the lifeblood of money away from the companies that would benefit from maintaining the energy status quo.

People become engaged when they are doing something they believe in, even more so when they start to make a living or have a financial return from these developments. It is not all that matters, but if these schemes have a financial benefit for the individuals involved, they see the value and impact of them in their own lives. The more people become engaged, the better, because once they are engaged, they start to fight for what they believe in. The fight in turn creates a movement, which gives the politicians the mandate they need to create the right conditions and to stand up to the big business tide that at this point inevitably flows the other way. We need a movement across the world, taking back the power. Our children need us to build this movement and to succeed in repowering our communities.

Our success will be a win-win, creating a new form of energy system that is held in common ownership. A new form of system that brings energy production much closer to home, creating jobs in the process. It is a system that will help individuals and communities control energy costs, yet keeps the money spent locally. Importantly it is a system that will help our communities to take the steps towards zero carbon as we build it. So let's set off together on that journey. The journey where we perhaps start with one small step to prove the concept to ourselves and our community. But where steadily we increase our resilience, reduce our carbon emissions and create energy for our community.

This section of the book is designed to be your guide and companion, explaining some of what you will need to consider in order to take the first steps to repowering your community. I realise that communities differ and have vastly different needs, and that legal frameworks, tax structures, incentive schemes and local conditions are as diverse as the places on earth. Accordingly, I have attempted to write a guide to the process that covers the broad principles, rather than local details. I will no doubt have missed many important nuances in your case, and I am sorry not to be able to write a guide that covers it all. I have to confess it is mainly focused on setting up a generation business, and predominantly electricity focused, which is of course only a small part of the puzzle. Many of the principles are transferable to the other potential business streams. However, I hope that this section is a good enough signpost to get you started, and gives you a plan and an order in which to proceed.

It covers what I feel are the essentials:

- People and team
- Business possibilities
- Finding a site
- Technology
- Permissions and legal aspects
- Financial and business planning
- Promotion and marketing
- Project management
- Building a sustainable future

So I encourage you to step up and become an energy revolutionary. As social entrepreneur, supporter, investor, farmer, municipality – you tell me where you want to be in this story. Let's not leave it to the experts any more. If you have no work, create a business for yourself in this area. If you have savings put them to work on the solutions in your town. If you have skills to offer, you are needed here. If you have wondered how to do something useful about climate and energy, now is the moment. It's time to step forwards with courage and take action. Here is one way.

CHAPTER 25

PEOPLE

We're not on our journey to save the world but to save ourselves. But in doing that you save the world. The influence of a vital person vitalizes.

Joseph Campbell, *Journal of a Nobody*

Your call to adventure.

If you have picked up this book, you have taken a step to becoming part of the vital journey of transforming the energy system. If you have picked up this book, in some way you have been called to join this adventure. Whatever your skills and experience, background and outlook, there will be a role for you in the process of making this change. There is a global movement of individuals taking action in their own local area, and you are now part of that movement. What part will you play? What gifts will you bring?

If you have read this far, you now understand some of the barriers to change, a little of what is possible, and some of what has been achieved elsewhere. Hearing the stories from other communities, it seems there is no technical reason why we cannot make this change. Most of the stories that we have heard start with a few people's vision and commitment to building the solutions in their local area. At some point each of those people just decided to get stuck in and make something happen, despite the challenges.

We are constantly bombarded with conflicting messages about our ability to change things – particularly in relation to energy and climate change. The problems seem so huge that it is easy to be left with the feelings that they are unsolvable. Feelings of powerlessness lead to depression and hopelessness, and most people choose to tune out the awareness and avoid those feelings. Another common response is to point the finger at others, like the government, and say that they are not doing enough. These huge, seemingly unsolvable topics can be depressing, and truly governments should be doing more, but this does not mean that we should not take action ourselves – in fact quite the opposite. We are faced with problems so large and complex that governments will struggle to find pathways to solving them. Our governments are framed in a system where corporate interest has become a large driving force in policy making, and where the ability to achieve the systemic change we need seems severely limited. My feeling is that the only way we will solve these issues is for enough people to get involved with creating the solutions, and to start doing so in their own community.

> *We are the ones we have been waiting for.*
>
> Alice Walker

However hard the changes we need may seem from where we sit now, they are definitely possible now, as others have shown us. These changes start with individuals who stand up and put themselves forwards for the challenge. That's you and me. We all have a choice in life about how we respond to these challenges – either with fear or with courage. Fear inevitably keeps us where we are in the status quo, whereas courage has the potential to transform. Courage means taking risks and putting yourself forwards when you are not certain whether you will really succeed. Courage means stepping out on a path whose route we do not know, but stepping out anyway. It is a well-travelled path that throughout history has led our societies forward. Stepping out into the unknown to go and seek a new vision of how our society could be, this is exactly what is required now. Some of the pioneers have taken the hardest steps, shown that there is a route to bringing our society back into balance with nature in terms of our energy use, and essentially that it is possible now. But many more people need to get involved to make this a mass movement that achieves a total transformation. It is time to 'birth' a new way of being on this planet, one that no longer relies on fire as the prime source of energy, one that brings our species and our societies back into harmony with the patterns of nature.

> *We have it in our power to begin the world all over again. A situation similar to the present hath not appeared since the days of Noah until now. The birthday of a new world is at hand.*
>
> Tom Paine, *Common Sense*, 1775

So it is time to act with courage and step forward to lead, in our own way, in our own community. The power of one person stepping forward is not to be underestimated; in fact it seems all change starts with that. When you look at all great movements for change throughout history, they often gained form, direction and momentum from one person's action. Who would have thought that the 42-year-old seamstress Rosa Parks' refusal to give up her seat on a segregated bus in Montgomery in 1955 would have been the act that propelled the US civil rights movement forward so dramatically? Yet this is exactly what happened, and her action became the start of the 381-day mass boycott of the buses that eventually ended in change to the laws of segregation on buses. The iconic act of one individual opened the doorway for change to begin. Her spontaneous action that day was, of course, informed by her involvement in civil rights issues and the fact that she lived in a time of great injustice.

She said later, "When I made that decision, I knew that I had the strength of my ancestors with me."[1]

Whilst energy and climate issues are very different from the struggles of human rights activists in 1950s America, the principles are surely the same: that if change is to happen, it must start somewhere, with someone acting as a determined focal point, as the people in the stories we have heard have been. Without someone acting as that focal point for their community, change is simply impossible. Our children and grandchildren need us to find the courage and determination to step into this place of possibility where change can happen. The moment we cross this threshold, undoubtedly we will be also be able to draw on the strength of our ancestors as we create a future for humans on Earth.

So this is our chance. It is time to step forward with courage and determination. It is time to organise in our town, it is time to build bridges in our community, it is time to develop the solutions that we need, and lead something we believe in. At this point, what have we to lose? If we resolve to get involved and to step into leadership now, things will move around us in unseen ways. The moment we are committed to the path, opportunity will begin to present itself, and the adventure will start.

Stepping into a place of leadership need not embody the top-down leadership styles of old, in fact it is essential that it is done differently. As defined in one of my favourite books on the subject, *Synchronicity* by Joseph Jaworski:

"Leadership is all about the release of human possibilities. One of the central requirements for good leadership is the capacity to inspire the people in the group: to move them and encourage them and pull them into the activity, and to help them get centered and focused."

This requirement for leadership that inspires is needed within your team, and can come from all members of the team, and will travel out into the wider community. Inspiring people is all about communication, telling the stories of hope and change, acknowledging that we all have an important part to play, that we all matter. It's time to tell inspiring stories, and then to lead in making them become manifest.

Is this your moment to step forward to lead? What would you have to let go of to get involved and step into this exciting space of creation? What will you be leading for? What might you find along the way? How might it feel to look back on the challenges and your eventual success? Prepare yourself for an adventure that will need all of your determination and creativity.

Where you stumble and fall, there you will find gold.

Joseph Campbell

---

[1]   The Story Behind the Bus – The Henry Ford Museum, www.thehenryford.org/exhibits/rosaparks/story.asp

This adventure will not be without challenges and sacrifices, and many of the pioneers I have spoken to in writing this book have expressed both the immense sense of fulfilment of having achieved the end result, and the personal challenge of the journey, and the sacrifices they have made in the process. When I first set off on my journey in this area 15 years ago, opportunity did present itself, but our plan failed and ended in dispute, so I was disappointed and put off. However, when I sat in a room in a pub in Lewes with the Transition energy group 10 years later, I knew that the time had come for the idea I had been carrying all that time to manifest. At that moment all it required was for me to step forward with determination for it to start becoming real, so that's what I did.

If you step forward, in whatever way, to represent the change you wish to see, others who are also seeking that change will be drawn to you. That is what happened to me when I went about setting up Ovesco. I was in a group with a range of people, some of whom became my fellow directors. Between us we had the diverse set of skills, experiences and outlooks that we would need to pull off our first projects.

It was the start of the adventure. Without a map to guide us we set off along a challenging path with many moments where it seemed that what we were trying to achieve was simply going to be impossible, and yet we still achieved our vision. You too can do the same.

A couple of years ago I had the privilege of being invited to speak at the launch of Brixton Energy Project's first share offer, for the first community energy project in inner city London. The passionate team who developed the project told the 50 or so people assembled about the details of the scheme, the investment opportunity and why it was a great thing to be a part of. There was a palpable sense of excitement in the room. I felt very humbled to be there speaking that evening. As I stood up to take the floor, I encouraged everyone to pause for a moment, to look around at their neighbours in the room, and to truly drink in this moment. "This is a night you will remember for the rest of your life," I said. "This is a night you will tell your grandchildren about, the night when we came together to build the first project in London that was the start of a big change, the start of the process of re-democratising our energy system and bringing it back into balance with nature. Whilst it may seem like a small step, it is nonetheless a significant step, the effects of which will be felt for many years to come."

So take a moment and reflect on how you could play your part in this. Whether this is all new to you, or you work in an energy utility right now, there is a part for you in this story. Your skills, passion and creativity are needed.

What are the stories that you want to share with your grandchildren? It could be that you are ready to lead the formation of a new company in your town. However, there are many other ways to get involved. Perhaps there is a group in your community already considering this that you could help with one of the many tasks:

- Finding sites
- Promoting the idea
- Doing the finances
- Facilitating great meetings
- Bringing people together
- Organising
- Investing and encouraging your friends and family to do the same

We will explore the many different things that will be needed to make a first project happen in later chapters.

ASSEMBLING A TEAM

Of course this is work we will not be able to complete on our own. Assembling a team around us to share the work and the journey will be essential. You may have people in mind with whom you have worked that you can call on, or maybe you are in a local Transition group where there are other interested people to work with. Great.

Getting a clear vision and setting a clear intention will be an important part of galvanising a group and supporters to action. Sometimes one person with a clear vision, clearly expressed, is all that is needed to give people the comfort to get involved. At other times a vision will have to be borne out of a group coming together and weighing up ideas until they have set a point to aim for. That could be a whole range of things, from the small to the all encompassing:

- Achieve the first community-owned renewable energy project built in our town
- Generate 20% of our town's energy from renewable sources within a year
- Become a net exporter of clean energy
- Eliminate fuel poverty
- Create a new social business
- Create local sustainable employment
- Create a new form of bank based on local renewable energy
- Generate all our town's energy by 2020
- Energy revolution!

What is it you want to achieve? When do you want to achieve it by? Who could take the actions needed? What skills will you need to achieve this vision? Where will it happen? Get the vision clear, and people will be able to join you. If you can, break the big ask down into chunks that you can set to a timeline.

There are many different ways to clarify a shared vision with a diverse group of people. Perhaps you will call on the help of someone in your community who is an expert in facilitation and holding constructive meetings to help carry it out. Techniques like 'open space'[2] can provide the freedom for a diverse group to come together and share their ideas in an open forum, and then, collectively, set priorities and next steps – defining purpose and aims.

If we are leading the process of setting up a group, a number of factors should be borne in mind when considering who to work with and who to invite in. We may be embarking on the course of setting up a social business, with all the legal responsibilities that that entails. It may involve creating a job for yourself or for others involved in the process. If we are successful, we may raise money from our local communities, spend it on a renewable energy asset that generates a financial return and share those returns with our investors. In time we will be able to sell that electricity or heat to people in our community. First we need a team of people willing to be legally responsible for all aspects of running this new enterprise. Second, we need a team with a broad breadth of experience and skills to cover the many bases required to make a business successful. As you will read in the coming chapters, these range from technical to legal and financial, from marketing and promotion to investor and project management. We don't necessarily need people with all those skills in-house, but ideally we need a core team with some experience in this wide spectrum, so that when experts are brought in, advice given can be understood and used effectively.

Perhaps most important of all, we need a core team around us who are willing to commit to the process and see it through, a team of people who are used to getting things done, and above all a team with lots of enthusiasm. In selecting this team it is important to recognise that this path will not work for everyone. Whilst it will be essential to have the support of all the interested people, it may not be appropriate to have all of them as our key team members.

In setting up Ovesco I saw the opportunity to establish the company and bid to run a grant scheme distributing grants for renewable energy projects on behalf of our local council. We had a very limited time to set up a company and write a convincing bid for this work. The Transition Town Energy Group were meeting and I was chairing the group at the time, and when faced with this opportunity I had a difficult decision to make. Rather than trying to bring the whole group along in the process of establishing this new company, I decided to get on with it and then approach some of the members of the group who I felt had the right set of skills and attitude, and the personal motivation to win this bid. I contacted them, and we quietly got on with the bid away from the main group. In doing this I rather upset others in the group, but I felt it was essential to have a smaller, more dynamic group of people working on the idea if it were to have a chance of success. With hindsight I could have done this

---

2   Open Space Technology, http://en.wikipedia.org/wiki/Open_Space_Technology

with more tact and diplomacy, but we did go on to win the bid, and that was the start of our journey.

Recently I was giving a talk in a village hall to another group looking at what they could do to power their own town – one of the first towns that was threatened with fracking in the UK, Balcombe. At the end of the talks explaining the options, people gathered into groups to discuss the local opportunities they could think of and the part they might be interested in playing. As the groups came back together to feed back their ideas, there was a buzz of excitement in the room. It seemed a great moment to help this community take their first step to forming their own community power company, so I asked all of those there who supported the idea to identify themselves, and was greeted by a pretty unanimous show of hands. I then asked who would be up for acting as a director of the company that would be set up and taking legal responsibility for it. At this point only about five hands went up. Most people were supportive but didn't want to be legally involved, but the group that did were the obvious core group. This new group went on, within a couple of months, to become the directors of a new company – Repower Balcombe. In this case the team that had formed spontaneously had an excellent mix of skills and experience that would carry them speedily through the first stages of their set-up with minimal issues.

So if you find yourself assembling a group to take on this challenge, think carefully about who to pick to join you, and how you frame the process. Pick people who will create a dynamic group with diverse experience and who will inspire and challenge each other. Fundamentally, pick people who share the desire to succeed in the mission, people who will not stop until you have succeeded. There will be plenty of moments where you feel severely tested, so having people who are willing to commit themselves and have a passion for the task is essential. Once a team starts to assemble, the creative process really starts.

ADVISORS AND EXPERTS

You will need to draw on the skills of quite a few specialists in the process of getting from an idea to a completed project. You may have some of the skills you need in the team that has come together, but undoubtedly there will be some skills that you need to make the project a success that you don't have.

You will need experience and expertise in the following areas to make your project a success:

- Renewable energy technology
- Project management
- Legal – leases, contracts and legal structures
- Financial – business planning, financial modelling, tax and accounting

- Promotion, marketing and public relations
- Community engagement and consultation

It is highly likely that most of these skills exist in our community, and that the individuals who have them would be supportive of our plan. If they are, there is an opportunity to get people involved initially on a *pro bono* basis as you get your first project under way, with fees either written in on success, or promised on the next project. In my experience people are often happy to help with these schemes, excited by the thought of doing something good for their own town. So ask and you might be pleasantly surprised!

ENGAGING THE WIDER COMMUNITY

When to start engaging the wider community will depend on the complexity and potential impact of your project on people's lives. If your project is fairly simple and not going to interfere with the running of most people's homes and businesses, you probably want to have it ready to go before you start letting people know too much about it. Generating a lot of interest in the idea before you have a project that is ready to go live could result in a false start, particularly when you are going out asking for investment. When you announce the project, you really want to be in a position to accept investment, since the announcement will be a prime moment for media attention and therefore public interest. These are things that need to be capitalised on to ensure you raise the investment needed in the appropriate time.

If you are preparing a complex project such as a district heating scheme, the consultation process will inevitably start early on – but even then you want to be presenting a clear and well-constructed proposal to get people's trust and support in making the project happen.

Support from your local community will be critical to delivering a successful project and business, so thinking about how best to engage with people in an open way will be essential along the way. Selling the benefits to them personally will be an essential part of this. Make it personal. Engaging people who stand to get work from the project will be one part of this, and is essential. Selling the benefits of lower bills or security of energy supply is perhaps going to be an easy sell in many places, tangible benefits to people's lives that will make your projects fly. Offering an opportunity for people to invest in an asset that is doing good in their local community whilst making a reasonable return is surely a win-win!

HARDWIRED FOR CO-OPERATION

I love Frances Moore Lappe's analysis of human nature in her brilliant book, *Getting a Grip*, and think it is helpful to all embarking on this path. She drew out four key strands:

- We are hardwired for co-operation – studies and analysis have shown that essentially we have evolved as a co-operative species that we got here from days as tribal creatures to modern times through co-operation. Neuroscience has shown that our brains are wired to make us feel good when we are co-operating with others.
- A sense of fairness lives within most of us – we have learned that injustice destroys the community on which we are dependent.
- We are problem solvers – we have become the dominant species on our planet because we are problem solvers and 'doers'. We like to make things happen.
- We are creatures of meaning – we want to have value beyond our own survival, to be good ancestors, to make the lot of our children better, to leave a better world.

Most of us essentially want our community to thrive and succeed, but perhaps don't know what to do to address our concerns about energy and climate change. If you manage to start a project in your town, however small at first, it will come to represent the growth of the new way, and will definitely gain momentum and support as it goes. You will build a beacon for hope that others can share in; that's got to be a good place to start.

CHAPTER 26

BUSINESS POSSIBILITIES

Having spoken to people all over the world about the community energy enterprises that exist today, there seems to be a range of common themes as to what a community-owned renewable energy business could look like. In this section I have tried to give a very basic outline as to what I think those possibilities are. Perhaps you start with just one of these options, as we have done with Ovesco, or it may be that you tackle a number of them at once. I go into detail of many aspects of what is involved in later chapters so if you don't understand any of the terms below they will be explained as we go on.

GENERATION BUSINESS

Setting up a generation business involves finding sites to locate and build renewable energy generation projects. As a generation business you develop the permits to build the project, and then raise the cash needed to fund the construction. This might involve raising community investment with commercial debt to fund the construction. Once fundraising is complete, the plant is constructed. The operational plant makes a return on investment by collecting any feed-in tariffs or other incentives and by selling the electricity generated. The company manages the asset moving forwards and ensures it is running smoothly and therefore generating energy and a return for investors. The company may go on to develop a portfolio of generation projects across a range of technologies and locations.

Example: Ovesco

SUPPLY BUSINESS

A supply business is set up to sell energy directly to individual consumers or businesses. In some territories this is simply not possible because of regulatory issues. In others, communities have managed this very successfully as we have heard in some of the earlier stories. In many countries, to sell electricity direct to homes or businesses requires a licence or permit. Applications for these licences are often arduous tasks designed for large corporate entities with big balance sheets and a large staff team. As part of the development of this business an energy supply business may need to be established to supply heat through a district heating system, or electricity through a micro-grid or island grid. In some circumstances an energy supply business is also needed to sell the output from the generation business described above. For some projects a distribution network will need to be developed in order to facilitate it, to deliver the energy from the place where it is being generated to where it is needed – such as happens with new build district heating systems and island grid systems in remote areas. Running the business may involve building and managing the network, the billing of the end customers and balancing the system in terms of generation supply. In many cases it will involve adhering to local energy supply legislation, and this may be a complex process. Setting up the company could involve raising funds from the wider community as well as debt finance to fund the construction of any new distribution network required, or the purchase of an existing network.

Example: EWS Schönau

ESCO BUSINESS

ESCOs or Energy Services Companies sell energy as a service. A simple way to think of this is as follows: Right now if you want light you buy a lightbulb and plug it in, paying your electricity bill for the energy you use. An ESCO would supply you with light and manage the cost of buying the lightbulb, as well as paying the energy bill, in return for a payment for the light that you use. If you run a business with a range of demands for energy to deliver your product or service, then an ESCO might be able to provide you with that energy requirement for less cost than you have traditionally paid. This could be achieved by making the appliances or building you use more efficient – like swapping standard lighting for LED lighting. It might also be achieved by generating some of your energy needs on site.

Establishing an ESCO to provide energy services is potentially a complex process requiring a wide base of knowledge and skills. Most simply put, it would involve identifying projects where energy savings could be achieved, and quantifying the investment required to achieve those savings. If the host site was keen to proceed, the next phase would involve funding the cost of the energy saving equipment and/or renewable generation equipment, project managing its deployment, managing its ongoing operation as well as billing the host site for the energy saved or generated. The costs would be charged back over a number of years to recover the capital investment and pay a return on that investment. Generally this fee is set so that the host makes savings when compared to the cost of the energy they had been importing from month one. After the capital investment has been paid off, the systems installed sometimes become the property of the host business, therefore locking in those energy savings and cost reductions moving forwards. In other cases our ESCO retains ownership of the energy saving technology and generation technology and becomes an energy partner for the host moving forwards. Setting up the company may involve raising funds from the wider community as well as debt finance, with a return on investment being paid as the savings are made and bills are paid.

COLLECTIVE PURCHASING OR 'AGGREGATION'

Many communities develop schemes to purchase energy or energy saving and generation devices collectively. Some communities do so on an informal basis where a group of homeowners come together to buy solar across all of their homes in one go. Others do so in a structured way, with the same outcome being achieved but using a co-operative structure to formalise the process. There are many things that can be purchased in this way from electricity and gas to energy saving or renewable energy devices. There are even cases of solar energy businesses joining together to collectively purchase the products they sell and thus achieve economies of scale. All of these schemes work on the principle that we are better off working together, and that economy of scale is achieved if we do. There may be a case for forming a business around this idea or using this as the starting point for your business. It may be easier to get together with your neighbours and look at what product or service you all need, and then see if you can leverage your collective buying power to help all access them.

In some communities collective purchasing, also known as aggregation, is used as the first step in the development of local renewable energy assets. By aggregating local power demand, the community has the power to issue long-term energy purchase agreements. This is in turn can be used to drive supply from renewable energy sources, and therefore their development locally. Community aggregation can also be used to deliver bill savings to local residents as well as giving choice in often monopolised energy markets. Communities or municipalities following this route can also help their clients reduce their consumption by implementing energy efficiency schemes.

Example: Marin County CCA

CHAPTER 27

LOCATION, LOCATION, LOCATION

It's all about location – your location.

> *Take the first step in faith, you don't have to see the whole staircase, just take the first step.*
>
> Martin Luther King Jr.

If you are planning to start with generation as your first priority, you cannot have a renewable energy project without a site on which to locate the technology. As a start-up and a community group, you are highly unlikely to own a site that could host your first project. However, that is not a problem when you know what to look for and how to present your ideas and plans to others in your community.

Wherever you live in the world, there will be sites around that can be used to generate power from nature's abundance; your first mission is to find a few of them for your first project. If you are based in a town or city, there will almost certainly be a large warehouse or retail outlet in one neighbourhood that could be potential hosts. If you are living in a rural community, then farmland and buildings are ideal locations. The various different technologies that community companies can use to build their own power stations need very different locations and conditions, and I have spent some time looking at the needs of each technology in the next section.

There are a few common themes that will make a site work, whatever the technology:

- Access to the primary resource. For instance – the basics: if you are planning to put solar panels on a roof, then it must be a good, unshaded roof. If you are planning a biomass project, you need a source of fuel to run it.
- A grid/network connection and/or an energy consumer to buy the power or heat that you are producing.
- A site owner who is willing to lease the area where the technology will be located for the term of the project – which can be up to 25 years.

- Support from your local community.

RESOURCE ANALYSIS

The starting point in your search for your first site is almost certainly to carry out a resource analysis of the renewable energy potential in your area. Whether you are urban or rural will affect what is possible, and recognising that fact will make your task easier. In the section on technologies, I lay out in some detail what to look for when considering the various technologies, and what might be appropriate. I have installed the full range of renewable technologies over the last 10 years, and found only one, solar PV, that is easily applicable in pretty much any situation. I strongly recommend that new groups focus on this technology for their first project, because it is by far the simplest and least obtrusive to get set up and keep running. It is also quick to get moving with. That said, many groups have been successful in using other technologies for their first project, so it may not be the best option for you and your community. If you have a river running through your town with a derelict mill, you may naturally be led to look at the opportunity of using hydropower first.

If you want to look at the primary resource available and being used in your area, there are a few obvious things to do.

The first is to look at what other technologies you can see around you. Are there wind turbines or solar panels on local roofs? Or your local environment – are there woodlands in your vicinity, or does a river run through your town? Careful study will lead you to other users of renewable energy and, with a little bit of work, you can find out exactly how the technology is performing in that situation.

In the absence of local clues, the second option is to look at local or global resource maps, which will provide you with an overview of the potential for the various technologies, wherever you are in the world. There is a very useful online tool, the Global Atlas for Wind and Solar, which has been developed by the International Renewable Energy Agency (IRENA). This is a comprehensive information platform showing the potential of renewable forms of energy. It provides resource maps from leading technical institutes worldwide, and tools for evaluating the technical potential of renewable energies. It is fully searchable, and will provide you with the average solar radiation values and windspeeds for your location. There will almost certainly be local versions of these maps for your area.

The third, and probably the easiest, option for assessing the resources available to you is to seek out and speak to your local experts in the field of renewable energy. Those companies and individuals that are building or owning systems in your location will be able to tell you exactly what they are capable of and where to start. You may even find fellow enthusiasts who want to join you on the journey.

GRID CONNECTION / ENERGY CONSUMER

If you are going to generate energy you need someone to sell it to if the project is to work financially and offer a return on investment. If you are making electricity, you must be able to plug it into an existing grid connection or local network, or have someone who wants to use all the power you will generate. If you are making heat, you need to be close to someone who needs that heat, and have a method of transporting the heat from place to place.

I will go into the details of what the various technical issues are and how to meet them later but to whet your appetite here are a couple of great projects:

DRAKE LANDING SOLAR COMMUNITY, ALBERTA, CANADA

The Drake Landing Solar Community is located in Okotoks, Alberta, 15 minutes south of Calgary. Ninety percent of space heating needs for the community's 52 single-detached homes are met by solar thermal energy, a feat unprecedented anywhere else in the world. It achieves this through a combination of solar thermal panels located on the roofs of the garages, district heating pipes connecting the solar panels up to all the houses, as well as to an energy centre, which in itself connects to a seasonal solar thermal energy store. This is the first major implementation in North America of seasonal solar thermal energy storage. Solar thermal energy is collected in the summer, stored underground, and then returned to the homes as heat during the winter.

To store the solar energy they are using borehole thermal energy storage (BTES). It is such a great concept – a BTES system is an underground structure for storing large quantities of solar heat collected in summer for use later in winter. It is basically a large, underground heat exchanger. It consists of an array of 144 boreholes drilled to a depth of 35m and about 2m apart. After drilling, a plastic pipe with a 'U' bend at the bottom is inserted down the borehole. To provide good thermal contact with the surrounding soil, the borehole is then filled with a high thermal conductivity grouting material. During the summer when excess heat is being produced it is pumped into this network of underground pipes, where it heats up the rocks below the ground. In the winter when heat is needed it is pumped back out again.

It's a great story, and there is even an app available on their website that shows you exactly what is going on with the energy flows in the system. Amazing to see how warm the system is even in the depths of winter![3]

---

[3]   Drake Landing Solar Community, www.dlsc.ca/app.htm

THE ISLE OF EIGG, SCOTLAND

Before February 2008 the people who live on the Isle of Eigg, a remote Hebridean Island 10 miles off the Scottish coast, only accessed power from a collection of noisy generators. Each home and business was responsible for its own power and it was not only expensive to run but unreliable.

On 1st of February 2008 after much hard work from the islanders and a whole load of other contractors, the generators fell silent as the islanders switched on a renewably powered electricity grid. For the community of the Isle of Eigg this project was the culmination of 10 years' work after buying the island in 1997.

The system they have built combines hydro, wind, solar and batteries to provide continuous power for all of the residents and businesses. They have one 100kW hydro turbine, two smaller hydro turbines at 10kW and 9kW, four 6kW wind turbines and 50kW of solar PV generation. They also have some diesel generators that provide backup power but the renewable generators actually provide over 95% of the electricity they need.

The power is distributed around the island by 11km of buried high voltage cable, connecting both the generators and households and businesses that use the energy. Homeowners are limited to 5kW of peak power consumption, and businesses 10kW and this means that people have to be conscious of what appliances they run. People are kept up to date with the status of the system and how much power is available by regular postings in public spaces, so they know if the power is running low. Residents and business users buy pre-paid electricity cards to be able to access the power. This reduces credit risk for the company established to run the system on behalf of the community, Eigg Electric Ltd. Rates are set at levels that allow the company to fulfil its commitments, but that are considerably cheaper than they were paying when each resident was running their own generator. A team of islanders has been trained up to maintain the system, and the islanders worked with a range of contractors to make this groundbreaking project happen. This could be a blueprint for isolated communities around the world.[4]

HOST SITE

In many ways this final point will probably be the most crucial part in establishing the right site. You will need to find a person or company that is willing to take a risk with your unproven company, and to sign a lease that ties them in for perhaps 25 years. This is something that can get tricky. When you then tell them not only what you want to do, but also that you are going to raise the money for completing the project from a share offer in the local

---

4   Green Eigg, http://islandsgoinggreen.org/

community, you will no doubt get some long looks and raised eyebrows!

When my fellow Ovesco directors and I approached our first host site we were met with a fair degree of scepticism. If you can imagine the scene, we managed to get a meeting with the directors of our local and much loved brewery – Harveys of Lewes. We were sat in their very old and rather wonky boardroom, amongst the trophies, very old photos of their forebears and stuffed fishes in display cases. Harveys is a traditional old English family brewery in the market town of Lewes in southern England. It is a very green and community-minded company, which really gave us a bit more confidence in our approach. As we sat and explained what we hoped to do, we were lucky to be met with a very positive response – they really liked the idea. But, as business people, they needed to be convinced of the business case for doing the project. They heard our pitch and went on to ask some very good questions about what we were proposing, the following among them. Would they be better off doing it themselves? What are the implications and risks for their business? What happens if we don't raise the money? How does the technology work? What would it do for them? What were the conditions of the lease? What if our company failed?

Hamish Elder, one of the directors, talked to me about the experience of working with Ovesco:

> It's a highly technical specialised field for a start so at first what you were greeted with was a whole barrier of ignorance. There was that hurdle to break through. We have been around for 200 years so pretty much everything we do has an element of radical change to it. We don't dismiss anything now at first glance. You have to hear out projects before you decide whether they are for you. We had the whole deal surveyed by a technical specialist to make sure that it would work.

The fact that there were so many questions seemed a very good sign, but it also highlighted that we had much to do to make the case convincingly enough for these busy people to give it the time required to consider it seriously. Our project was relatively simple in terms of the development process and what we were proposing to do, but it still needed a considerable amount of research in order for us to present a compelling case. We then went off to begin the process.

This was as much a process for us, our potential business, as well as for our hosts and future investors. For our prospective landlord and project partners, we pulled all this into a document that laid out the proposal in detail. It ran to more than 10 pages and covered all aspects of the project: about Ovesco, who we were and why we were doing this, about the technology (in this case solar electric PV), the market for PV and the income streams, what we were planning to do on Harveys' roof, ownership and funding, lease arrangements and planning permission, electricity supply arrangements, how the installation would be carried out and by whom, the benefits for our host, and, last but not least, how long we expected everything would take.

The process of getting the final go-ahead for our project was relatively short – about nine months from start to finish – and as it turned out we couldn't have picked a better partner to work with.

Hamish Elder continues:

> It did seem intuitively to tick all the boxes, because I personally couldn't really see a downside. The elements of upside were mostly PR related and that was difficult to quantify.
>
> It has been useful when taking people on brewery tours, to say by the way we have a 98kW solar power station on the roof, which is quite impressive as we are talking to most people about quite traditional things. It links well with our environmental credentials.

Who might be able to host your project?

- Groups that share your aims or have aligned missions who have land or buildings – e.g. local charities or environmental groups
- Local companies that have a large roof
- Large international chains
- Your local school
- Your local government
- Local farmers and landowners

It might be that in your hunt for a site you find a fellow traveller who shares the vision and is as keen as you to make the project happen.

Tom Parker is one of the directors of Repower Balcombe and has been looking for sites over the last six months. He shares the trials and tribulations of the process:

> I started looking for local sites for renewable energy in earnest last autumn when Repower Balcombe was forming. In truth I had been looking much longer than that, as the potential to provide all our electricity needs by renewables is barely scratching the surface at the moment. Having provided my family's electricity and renewable heat for some time, it just surprises me why others aren't doing the same. With this head start I already knew some really good sites for solar PV, not to mention wind and microhydro. A look at the local map quickly pinpointed some very large roofs on our patch, and a check on Google confirms the orientation and likely issues with shading.
>
> The big issue, however, is not so much the sites as who owns them and

whether you can get five minutes on the phone with them to try and get a meeting. This is where the group came in, and their contacts and ability to find the right person. Sometimes we got a speedy response, but more often than not it involves a long wait, as it is not necessarily going to be high up their priority list to call you back. Just when you think they are not interested, you get a call out the blue and you have few minutes to pitch your case and if you are lucky you have got a meeting, great!

Then you have got to do your homework, assess their site and listen to what they want. You know it's going well when they ask what part of the industry you work in, or come out with comments like 'so there are no downsides'. Alternatively, you know it's going badly when they just can't get the hang of the idea that you are doing this without being paid, and for the benefit of the community and the environment. The other potential hazard is that you sell the idea so well that they decide to put up the solar themselves. It is big plus for the environment, but little bit disappointing from the co-op's point of view.

Just when you think you are getting nowhere, you go off to buy your Christmas turkey from your local farm shop and think, why didn't you think of them before! A nice new south-facing barn, high electricity demand and best of all, friendly, straightforward owners who that find the idea of 30% less to pay on electricity easy to sign up too. Yippee, you have got your first site, and there are bound to be more to follow.

SUPPORT FROM YOUR COMMUNITY

If local people don't understand and support your plan, it will be very hard to make it happen. So part of the site-finding process is considering how the various parts of your community will view the proposition. Some of the stakeholders will perhaps be set to benefit from it in some way, and the more you can open that benefit up to all so that it supports local business and benefits multiple stakeholders, the better. It might be that if you were building a portfolio of projects and choosing sites that would directly benefit different groups in your community, even if they were not all optimal in terms of returns, they would be worth doing just to build a community of advocates around your project. There will no doubt be enough challenges without opposition from your neighbours. Making each project a win-win that accesses different resources, in the shape of both people and practical sites, is essential. In our work at Ovesco we have achieved this by installing systems on a wide range of buildings. A great way to create sites that touch a lot of people is to install PV on local schools, which benefit from reduced electricity costs as well as environmental education. This approach has helped Ovesco to make connections with a large section of the community; young people are powerful advocates!

CHAPTER 28

TECHNOLOGY

Let us put our minds together and see what life we can make for our children.

Sitting Bull

At this point I am convinced we have all of the technology we need to bring about the low-carbon transition of our society. It is now time to apply it and we need to do it fast. As you have heard in previous sections if we put our minds, hearts and creativity into the problem we can roll out the solutions in a very short space of time if we start in our own towns and communities.

Assessing what energy technology your community can use and where you could locate it really go hand in hand – it's the first step to seeing if you have a viable business. Your local environment will dictate what resources are available in terms of sun, wind, water, waste or biomass, and that will lead your enquiry. In this section we will look at what a good site looks like for each of the different renewable energy technologies available. We will explore what you can expect from each of the different technologies, a bit about their development and history and as well as what conditions they need to perform well. A badly sited, designed or installed energy generation technology will not give you the output and therefore the financial returns you are seeking to make your new business viable.

Probably the easiest way to quickly assess what technology will work for your first project is to find someone who is a specialist in the area. It may be that one of your team has this specialist knowledge that they can bring to the process to help speed your development along. In Ovesco we were lucky to have Nick as one of our directors, a very experienced engineer who had fitted a wide range of renewable energy technology to his house and had a very good understanding of what was possible technically in different situations, as well as myself with over 10 years' experience in building a range of renewable energy projects. This gave us a head start as we were quickly able to rule out many technologies as being unviable of very hard to deliver right at the start of our journey. In my experience of sitting in groups of people who are considering developing their own renewable energy project as a community for the first time, somebody has a 'eureka' moment and pipes up: "Why don't we just build a big wind turbine on the hill above the town – that would provide all of our energy needs." A great idea and one that may well be true, especially here in windy England. However developing and building a big wind turbine is a complex, costly, time-consuming and challenging process, so perhaps not always the best place to start for a fledgling company with no capital and no track record.

If you don't have the technical skills in your immediate group, someone in your wider network is bound to have some of these skills, so see if you can draw on them to assist. Most people involved in renewable energy are passionate about the subject – they have 'got the bug' and as such they love nothing more than helping others to get started too. Companies active in these areas can be called on to give advice as to what each technology can do for you. With some technologies this can be a fairly simple procedure, accomplished with a simple site visit or even a desktop analysis and with others it may require considerable upfront design and development work to assess if there is a viable resource. This assessment process can be costly and time consuming and does not always lead to straight answers. For your first project keeping it simple is likely to lead to success the quickest.

It may well be worth considering some form of partnership with a commercial company in the area you are looking at. A partnership could enable you to develop your ideas with specialist advice sat at the table, to avoid you making mistakes along the way. There are a range of formal and less formal ways that this could be structured and I look at this area later in the section on permissions and legal issues.

I am not going to go into lots of technical detail on all of the various technologies and how they work as there are many good books on the subject and there is no point repeating what is explained expertly elsewhere. What I am going to try and do is to give some insight into what the key parameters are that might make a business based on these technologies viable in the simplest terms. When we began our journey in Ovesco we decided to look at using our local river, the Ouse, as our source of power. We had seen some other groups' successes in this area using a 'run of river' hydro generator called an 'Archimedes screw', and had read the studies on the reliability of hydro as a resource. We thought this would be a great place to start, which of course it would be. However to develop a hydro project is a convoluted process, especially here in the UK, with multiple stakeholders and complex permissions required to access the resource. We Ovesco newbies spent about £10,000 carrying out studies on the River Ouse, looking at the potential of installing a hydro turbine at an old watermill site, only to find out much later that the authorities would not allow us to do it anyway! Tough lessons when you have limited resources to achieve your aim.

If you are a new group or coalition, your first project will be a test of your business model, as well as your resolve and determination as a group of social entrepreneurs. You will need to gain the trust of your community as well as their investment to make it possible. You may be starting with little or no resources to complete the development phase. It is therefore prudent to pick something as simple as possible, as there will be so much to do and so many little challenges to get from the great idea to a turning turbine or a generating solar panel. You don't want an overly complex technical project to stand in your way on the first go. For most groups who have yet to do a project but want to get started, solar PV often offers the simplest entrance point, and it is what I would encourage you to consider first. So let us start there.

SOLAR PHOTOVOLTAIC (PV)

> Unlike fossil fuel, solar produces no pollution and no miners get killed. Unlike nuclear fission, it produces no radioactive waste. It harnesses the power of the sun, which is the ultimate source of most energy on Earth. And it can strike the imagination of a people and therefore of their politicians.[5]

When I bought my first solar PV panel in the late 1990s I had no idea of the potential of the kit I was holding in my hands. Fifteen years on I have witnessed an amazing transformation in the costs, availability and deployment of the technology, in a journey that has become known to those in the business as the 'solarcoaster' for its many ups and downs.

Solar PV converts light into electricity and it is a technology with incredible potential. It has had an amazing story in its development to date. The technology is based on silicon, its abundance second only to oxygen in the Earth's crust. Sand, 30% by weight silicon, is reacted with carbon at high temperature in a furnace to create crude silicon and purified in various forms using processes developed for the creation of computer chips. The silicon ingots destined for solar panels are sliced into wafers. With the addition of tiny amounts of different elements top and bottom, each wafer becomes like a small battery when placed in the sun. So to increase the voltage each of these wafers is strung together and combined into a 'module'. Modules generally consist of a glass front plate, the wafers and then a protective layer to seal the whole thing together on the back; this sandwich is laminated at high temperature to seal the wafers away from water and to protect it from contamination. They are generally then 'framed' in aluminium frames to protect the edges of the laminate and enable them to be easily mounted. Modules are then combined together to form arrays, and their flying leads on the back of the modules continue the string. There are a few types of solar panel, mono-crystalline, poly-crystalline and amorphous silicon or thin film. They have different efficiencies and costs associated with them, with mono-crystalline cells generally being the most expensive and efficient (up to 21.5%), poly-crystalline being slightly cheaper and less efficient (up to 16.2%) and thin film being the cheapest and

5   'We Must Harness The Power Of The Sun', www.theguardian.com/commentisfree/2013/sep/29/climate-change-energy-sources-solar-power

least efficient (up to 9.6%). The different modules are better suited to different applications, and there are some obvious things that will guide which you pick – for instance if you have a small space available to install a system and you are trying to maximise your generation, then opting for thin film panels would probably not be a good idea! However thin film units weigh a lot less and may be useable on roofs that could not take heavier but more efficient panels.

In 1992 there was 0.38GW of the technology installed worldwide. It had arrived on Earth essentially as a by-product of the space programme which needed solar to power satellites – first commercially developed by Bell Laboratories for the space programme in the mid-1950s. By the end of 2014 the total installed capacity was around 178GW[6] – a massive jump in 22 years, and the speed of deployment is increasing. It is rapidly catching up with nuclear power with the almost static 375GW of installed capacity in 2014 of which, because of shutdowns and de-ratings, only about 276GW[7] is operational and for which costs are rapidly rising. Some European countries such as Germany, Spain and Italy have deployed huge amounts of photovoltaic technology, with Germany leading the charge in the early years. By July 2014, Germany had more than 37GW[8] of solar PV installed which provided just under 7% of its entire electricity consumption in the first half of 2014. The German people have invested heavily in the technology and have been responsible for the development of solar for the rest of the world as we have heard. In the UK it rose to over 5.4GW in 2014. During this period the cost of solar has dropped at a dramatic rate, to 5% of what it was 22 years ago and 0.5% of what it was in 1977.

More recently the focus of manufacturing and deployment has shifted away from Europe with China getting heavily involved in the manufacturing process and this has in turn reduced the cost of the modules dramatically.

Solar PV is a very versatile technology as it works equally well across a large range of scales – from the tiny cell in your calculator up to the 100MW plant covering many acres of land powering a whole town. This means it can be applied at an appropriate scale in your situation, from a local roof to a local field. Most systems are connected directly to the electricity grid, but it is also totally possible to connect PV systems to battery systems or island grids. People can start small, with one panel, and build as they need more energy.

To work well solar PV needs direct sunlight. Shade cast by objects onto the solar panels will stop them working; even if a very small amount of the surface of the panel is shaded, it can have a dramatic effect in reducing the production of the system. They can be placed on existing roof structures, or on frames on

---

[6] '40GW of PV Installed Worldwide as Europe's Renewables Overtook Nuclear in 'Tipping Point' Year', www.pv-tech.org/news/40gw_of_pv_installed_worldwide_as_europes_renewables_overtook_nuclear_in_ti

[7] www.iaea.org/pris

[8] 'FRAUNHOFER Electricity Production from Solar and Wind in Germany in 2014 July 21, 2014', www.webcitation.org/6RG21ExIC

the ground. They can be used in standalone devices such as the solar light, or integrated into a building to form the roof structure.

The solar electrical system consists of a number of components, the panels themselves, the mounting structure, the inverter, and the balance of system (cables, switches, meters etc.). The panels produce electricity as 'direct current', DC, similar to what comes out of a battery. Most electrical grids and appliances operate on 'alternating current', AC. The inverter is the bit of kit that converts the DC from the panels into AC for use in the building the panels are mounted on. Systems where there is no electricity network generally have a battery bank as a backup and to smooth the output from the system.

Where you are geographically in the world will determine how best to site the solar panels to get optimum yield. Where I am in the UK, siting your panels at a 30 degree tilt facing due south is the optimal position, but systems that face east or west are often viable; they just produce a bit less energy. If you are living on the equator your panels will be most efficient if they are installed close to a horizontal position. There are a number of computer simulation programmes available that can accurately model the output of a system design in a particular position; they will even take into account shading factors like nearby trees and distant hills, as well as using weather data. To start the process of accurately modelling a site we use a device called a 'suneye' which takes a fish eye lens photo from the site, and can then project the level of the sun and shading factors for a 12 month period from that one photo. This linked with the computer modelling gives a very clear picture of what a well designed and installed system will do in terms of energy production over the course of a year. Your specialist contractor or advisor should be able to help with some of these projections to feed into your business model. Less sophisticated programmes such as RetScreen[9] are available free and will enable you to work out yourself which prospects have a reasonable chance of being viable and are worth suggesting to your contractor or advisor.

If you are living in an area with an electrical grid, there is one other area to consider when looking for a site for your array of solar panels, and that is grid connection and the energy use on the host site. The process of connecting to your local distribution network may have costs involved, and it may also limit the size of system that you can install in a particular site. Normally there will be a process to go through to apply for connection of a new generation device; generally the smaller the system the easier it will probably be. In the UK for instance to connect home-sized systems up to about 4kW is effectively automatically permitted, but above that scale will need an approval which can take up to 60 days to determine. There are different codes regarding this in different parts of the world, once again, local specialists should be able to advise.

---

[9]   www.retscreen.net

Photovoltaic price and installed capacity

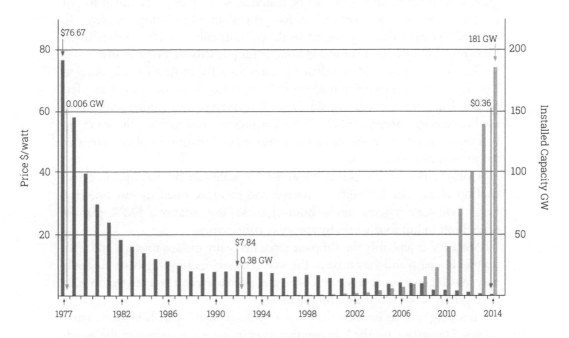

Source: Bloomberg, New Energy Finance & pv.energytrend.com, Earth Policy Institute

If you are intending to mount your system on the roof of an existing building you will need to ensure that the structure of the building is strong enough to carry the extra weight. A structural survey looking at the roof structure will determine whether the building can carry the extra loading – which is generally about 16kg per metre squared but can be about half this for thin film panels. It is also prudent to determine that the roof is in a good state of repair as your system will be in place for 25 years or more, and you don't really want to have to remove it to fix an old roof halfway through. Generally the easiest type of roof to install on is the metal sheet profile roof that is common on many industrial buildings worldwide. The most complex is often flat roof structures where wind issues and the inability to fix directly to the roof can make it tricky, but of course not impossible. Local conditions such as wind and snow loading and salt sea spray may need to be considered as may various conservation regulations.

Mounting solar PV modules on standalone frames on the ground can be a very good way to proceed if you do not have access to an appropriate roof, or you are looking to build a larger system. There are many of these 'field scale' solar projects in the world and they are probably the most cost-effective way to build solar PV at scale, as the construction process becomes very streamlined. Getting permission to construct these larger plants can be more complex and time consuming. Generally they use 'piles', pieces of metal driven into the ground to fix them down instead of concrete, so they are very quick to put up and have a much lower carbon footprint than when using concrete. To establish whether this is going to work will normally require a geotechnical survey to check whether the soil is suitable for pile driving on your site.

The development and permitting process for a PV project from looking at a potential location to being ready to build the project can be as short as a few weeks. This time is needed to do the various surveys and studies and put in applications for connection. On larger schemes this can take much longer, up to a year or more to get the various permissions and permits in place to be able to commence construction.

Construction of PV plants whatever the scale can be very quick. Small rooftop plants can be built, connected and commissioned in just one day. Large field scale systems can be built in weeks, for instance a 5MW plant is often built in just four weeks by specialist contractors.

Solar PV is probably the simplest place for many groups to start, and with no moving parts and very little in the way of servicing, running costs are really very low. It generates revenue as it generates electricity, and you sell this to your local grid or collect feed-in tariff payments.

Tracking systems increase the output by pointing the panels at the sun as it moves. Depending on the location they may increase the output of the panels from 10-35% with the greatest advantage for locations with a lot of cloudless days. However they are only worth fitting if the extra cost is not more than the value of the extra electricity generated. A tracker that increases the generated output by 15% is only worth it if the cost of buying and installing it is less than 15% of the cost of the panels it carries. They also have moving parts unlike the rest of a PV system, which adds a potential area of failure into the system that static panels don't suffer from.

Capital costs vary depending on where you are in the world and what scale you are building at. Outputs also depend on where in the world you are. In many parts of the world solar PV has reached 'grid parity', where it is cheaper to buy electricity from a solar plant than from any other form of generation. Diesel generated electricity will be significantly more expensive than supplying the electricity demand with a PV plant. Where replacement is not an option, supplementing a diesel plant with PV will dramatically reduce the costs of electricity supply.

Things to look for in planning a PV system:

• Grid connection – how much generation capacity can you connect to the local network? How much energy is required by the onsite consumer or host building?
• Structure for mounting the panels – is the system going to be mounted on a roof or on the ground? What does the structural or geotechnical survey say?

- Size of space available – solar PV takes up about 6.5m² for each kWp of panels if laid packed together on a roof or, when spaced out on a field, from about 11m² in the tropics to about 20m² at latitude 50 north or south and about 30m² at latitude 60.
- 1kWp will generate different amounts each year depending on the solar radiation in your area, more if you are in the equatorial regions and less if you are in the rainy north like me. But it still works well in all locations.
- Grid independent systems will need some sort of battery storage and in some cases diesel or other backup.
- Are there any good buildings in your neighbourhood that would fit solar panels that you could approach?
- Are there any friendly farmers who could be a host site for your solar array?
- For a field mounted system are there any environmental, planning or zoning restrictions?

SOLAR THERMAL

Solar thermal systems make heat out of sunshine. That heat can be used for heating things up and counterintuitively cooling them down also. Globally by the end of 2013 there was an installed capacity of 330GWth of solar thermal collectors in operation across 58 countries, which equates to a total of 471 million m² of collector. These systems produce an annual yield of 281TWh, which is equivalent to savings of 30.1 million tons of oil and 97.4 million tons of CO_2.[10]

China actually has the most installed solar thermal of any country in the world, nearly four times the capacity of all of the systems installed across Europe.

Predominantly, solar thermal technologies have been used for water heating needs, but increasingly solar thermal is installed to deliver space heating, industrial process heat, district heating and even solar cooling. I am not going to talk about concentrating solar power – where you put a field of mirrors in a hot place and reflect the sun's heat onto a tower, which gets very hot – as I don't suppose there are many community groups who will plan to build one of those at this stage!

Solar thermal systems for water heating are essentially pretty simple – as we all know if you put a black sheet of metal in the sun it gets hot. The panels are actually a bit more complex than that. The trick is to make sure nearly all the heat from the absorbed sunlight goes into the water to be heated and very little

---

[10] 'IEA-SCH Solar Heat Worldwide – Markets and Contributions to the Energy Supply 2012 Edition 2014 Published', www.estif.org/no_cache/news/single-news-item/archive/2014/july/article/iea-shc-solar-heat-worldwide-markets-and-contributions-to-the-energy-supply-2012-edition-2014-publ/

escapes back into atmosphere. The various types of panel differ in the ways they achieve this. For the better types of panel, the surface, instead of being black, has a 'selective surface'. This means that it absorbs energy well but limits the amount of energy it re-radiates and therefore would normally lose.

Generally the vacuum tube variety will extract a bit more heat per square metre of collector, so are good when space is limited or maximum heat extraction is required. The flat plate collectors are slightly less efficient, but also a bit simpler both in their construction and use. Essentially they both reduce heat losses by reducing conduction and convection of heat by placing the energy absorber in a vacuum – in the case of a vacuum tube system, or an insulated box in a flat plate system.

Both vacuum tube and glazed panels have tubes filled with flowing water to take the heat away from the absorber. Because water expands when it freezes and could burst the pipes, in countries where there are frosts, antifreeze is added to the water. This in turn means the water in the panels cannot be used directly for cooking, washing clothes or bathing and the heat must be transferred to clean water in a heat exchanger or a heating coil in a water tank.

Many systems that are deployed around the world, especially in China, Cyprus and Israel, use packaged systems where the tank is mounted horizontally above the collectors and natural convection used to transfer hotter water from the collectors up to the tank. When the water in the tank is nearly the same as in the collectors, convection stops naturally. Heat is extracted from the tank by water flowing in a heat coil in the tank with this water being driven around the normal water pressure. This eliminates the need for complex controls, sensors, pumps and an electrical connection, making the system cheaper and more reliable. Both glazed flat panels and vacuum tubes have been used as collectors in such systems.

In more complex systems, design to meet the loads has to be carefully undertaken and is often modelled with simulation software which will take into account the heat storage or load, the collector area, the orientation and the location to produce an accurate output estimate. Although shadowing will reduce the output of solar thermal systems, unlike PV systems the loss is only proportional to the lost sunlight. A television aerial pole that, in front of a PV system, could cause havoc, will have a tiny effect on a solar thermal system.

Having too much heat can be an issue so the system is often sized to meet peak load – or the point of highest demand – with a fraction of that being delivered at other times. Siting of the panels will affect their output, and your location will determine what angle will work best. In northern climes a 30 degrees due south orientation will yield the most energy – however if your scheme was for example delivering heat to a heating system, placing the panels on a vertical wall will give a flat line output that will be better suited to the need.

The heat is either used directly, in the case of a swimming pool or in some district heating systems, or is stored for use. The storage system can take a number of forms – from a simple hot water cylinder placed in your home, through to an 'injection well' ground based heat storage system like we heard about in the Drakes Landing Project. Hot water cylinders designed to store the heat for use in homes and businesses are fairly common with conventional heating systems. Generally systems that combine solar thermal with other traditional or renewable technologies use a store such as a hot water cylinder. For larger systems these stores also become larger. Increasing the size of the collectors and store may mean that the solar system can contribute to space heating requirements also. The 'solar fraction' of the heat energy required can be virtually 100% if the building is designed around this idea. Storage systems are at their most extreme in some district heating schemes. In those cases the storage takes the form of a huge insulated pit in the ground filled with water or a series of 'injection wells', which are heated by the solar thermal system, and then the heat is circulated around the houses and other users connected to the heat network.

Community-owned solar thermal systems might work when:

- Feeding a large heat user – such as a swimming pool, large building, or industrial process such as a water treatment facility
- Combined with other technologies such as heat pumps or biomass boilers to provide a total solution
- Plugged onto a district heating system
- Combined with some sort of inter-seasonal storage (be that a pit or injection well)
- And where heat can be easily metered and billed

Building a solar thermal project can be a relatively simple and fast process. Retrofitting a system onto a building with a high heat load for instance can be done in the course of a couple of weeks. Large-scale collectors are easily craned into place and fixed to the roof structure during the course of a day. The rest of the system, pipework, pump, controls and either heat exchangers or storage tanks can all be installed in a matter of days.

Careful design of the system is essential to ensure that the whole rig is optimally sized to meet the heat demand. An oversized system can become problematic as it can therefore start to overheat in summer months. Managing excess heat is not ideal or easy; systems are designed to cope with excess heat with various measures to prevent overheating and to keep systems safe in the event. There are two types of system design: pressurised or drain back. Pressurised systems are fully sealed, often rated up to 10 bar; they allow the heat transfer fluid inside them to get to high temperatures and pressures without flashing to steam and compromising the system, and have built in expansion vessels and safety discharge devices. Drain back systems empty the collector of heat transfer when the optimum temperature has been achieved or when there is not significant heat to collect. In both cases a control system will monitor the temperature at the collectors and the temperature at the bottom of the storage device and when the collector temperature is a few degrees higher it will start the pumps. Both designs will have an anti-freeze solution in the heat transfer fluid, which is normally water, to prevent freezing in colder climes.

When considering a solar thermal scheme you will need to investigate:

- Heat load of the host site
- Site for solar panels
- Structural strength if it is a roof-mounted system
- How the system will integrate into existing plumbing and heating systems
- Metering and billing of the heat delivered

Solar thermal systems can also be used for cooling applications, which on the face of it sounds paradoxical, but is actually very smart. The times when you need the cooling the most is when the sun is strongest and therefore the solar thermal panels are producing the most energy – a perfect match.

The usual method in use today is a closed cycle absorption chiller. Water vapour is cyclically absorbed and de-absorbed into and from a solution of ammonia or lithium bromide or silica gel. This is the same process that was used domestically in the early part of last century for gas fridges and is still used for some caravan and boat fridges. This method of using heat to produce cooling was pioneered by Albert Einstein in his spare time from formulating relativity.[11]

Solar heat can also be used to increase the efficiency of a standard compression cycle air conditioner. The heat increases the pressure of the refrigerant so that the compressor has less work to do.

Air can be cooled by allowing water to evaporate into it. This is how sweating cools the body. In places with a hot dry climate this allows cheap and simple low energy evaporative air conditioning, colloquially known as 'swamp coolers', to be used. The incoming air is blown over wet pads to evaporate the water. This is an open loop system in that water has to be constantly supplied. Such coolers do not however work well in humid climates.

Solar thermal cooling is still in its infancy globally, although there are many good reference sites already built. Invariably costs come down as knowledge and deployment rates go up. Generally the systems are bespoke and there are not yet pre-packaged systems available. However they have the potential to be an ideal technology in many parts of the world where cooling demand is very high and it is currently met using electricity.

---

[11] The Story of Einstein's Refrigerator, www.io9.com/5706535/the-story-of-einsteins-refrigerator

WIND

Wind energy has been used for over 4,000 years for powering our communities. Before that people sailed around the world using wind energy. It's pretty reliable stuff that has been harnessed for many uses around the world, from grinding corn to moving water.

The modern version of wind power sees the wind being put to work to make electricity. There are many different designs of turbines out there from the small home sized units to the large multi-megawatt commercial machines. The march of commercial wind power across the world has seen the installed capacity go up from 40,000MW globally in 2003 up to 360,000MW installed by 2014.[12] Close to that of nuclear generation.

In 2014 wind produced 39.1% of Denmark's electricity, 39.1% of the Netherland's and 9.3% of the UK's. In Spain, wind produced 21.1% of electricity in 2013 and 23.2% in the first half of 2014.[13]

Over the same period the costs of commercial wind energy systems have dropped dramatically with the world's wind power leader Demark reaching a point where wind is rapidly becoming the cheapest form of new electricity production. According to Denmark's Energy Agency, new wind plants coming online in 2016 will be generating electricity at half the cost of coal and natural gas.[14]

Small turbines and large commercial turbines all work well when sited properly but will obviously achieve very different levels of generation. There is not one right size of turbine out there as for some a very small amount of power can be totally transformational. Take the story of William Kankwamba who lived in rural Malawi and aged 14 built his own wind turbine. Using scraps and junk he managed to construct a turbine that powered four lights and a radio in his home. On hearing of his success people came from miles around to see his work – the first electricity in the area. He has gone on to write a book about his experiences called *The Boy Who Harnessed the Wind* and become a source of inspiration to many people in his community and around the world.[15]

My first foray into wind energy involved trying to build my own turbine from an old vehicle brakedrum, courtesy of excellent plans drawn up by an off-grid pioneer, Hugh Piggott of Scoraig Wind. I spent time hunting parts in scrap yards and winding the permanent magnet alternator that is the heart of this design. Sadly, having built the generator I never found a site to put the

---

12 'Global Installed Wind Power Capacity in 2014' , www.gwec.net/global-figures/graphs
13 'Wind Power Was Spain's Top Source of Electricity in 2013', www.theguardian.com/
    environment/2014/jan/06/wind-power-spain-electricity-2013
14 'Wind Power Undercuts Fossil Fuels To Become Cheapest Energy Source In Denmark', www.
    theclimategroup.org/what-we-do/news-and-blogs/wind-power-undercuts-fossil-fuels-to-become-
    cheapest-energy-source-in-denmark
15 Moving Windmills Project, www.movingwindmills.org

turbine up – but it was good learning to go through the process. Many people around the world have successfully built turbines following his excellent plans – now conveniently cooked into *A Wind Turbine Recipe Book*.[16] This guide along with others available on the internet are excellent pragmatic and practical guides which mean there is no reason why you could not do so also![17] In fact there is a movement of people all over the world constructing their own turbines in areas where access to power is not easy but can be transformational. It is amazing how even 50W of power can transform people's lives.

The other end of the wind turbine spectrum to these homemade machines is the Vestas V164 8MW which is, at the time of writing, the world's largest wind turbine. The blades are 80m long, and sweep an area of 2.1 hectares. Thirty-two of these giant machines are destined to be installed in Liverpool bay.[16,18] More commonly, commercial wind turbines range from a few kilowatts to a few megawatts in output. The larger the blades the greater the area of wind they sweep and the more power they can capture.

For wind turbines to function well they obviously need to be sited in a windy location and this needs to be free of other obstructions such as trees and buildings. Turbines generate well when they are in 'clean air'. As wind passes over buildings and trees it creates turbulence that is not good for production of energy or the longevity of the turbine installed. Most of us think that wherever we are is a very windy spot when the wind is blowing, but that doesn't always follow. There may well be places in your local area where wind has been used traditionally. Certainly in my area there are lots of old windmill sites dotted across the landscape where there are old mills or simply 'mill hills'. These are obviously going to be good potential sites. If there are obstructions to the airflow in the area they will cast a wind shadow, or an area of turbulent air for approximately three times the height of the obstruction. For small wind turbines, the whole of the swept area should be at least 10m above any obstruction within 100m.

Wind resources, as we touched upon earlier, have been mapped to different degrees in different territories and there are a range of different resources out there to determine the likely windspeed of your site. It might be there are simple banded wind maps of your location or more detailed mapping services that can predict likely average wind speeds in your area. The larger the project you are planning the more attention you need to pay to this area. To get accurate data it is common to erect an anemometer on a mast at the proposed site to the height you are planning the hub of your turbine to sit, and to leave it there for a year. Laser wind measuring devices (lidar) are expensive but have the advantage of not needing a mast and measuring at several heights at once.

---

[16]   Hugh Piggott's blog, www.scoraigwind.co.uk

[17]   Practical Action – Small-scale Wind Power, www.practicalaction.org/energy/small_scale_wind_power

[18]   Close up – Vestas V164-8.0 Nacelle and Hub, www.windpowermonthly.com/article/1211056/close---vestas-v164-80-nacelle-hub

Deployment and cost for US land-based wind 2008-2012

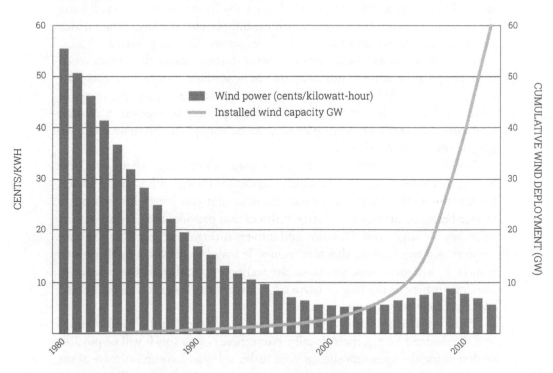

Source: US Department of Energy, http://energy.gov/sites/prod/files/2013/09/f2/200130917-revolution-now.pdf

In some places anemometer masts need planning permission but ground based lidar instruments do not. The data you gather will give you confidence in what output you will generate, and can be correlated to a nearby weather station which may have many years of data versus the one you have from your monitoring. Many airfields have wind speed data. You can then extrapolate an average expected annual output from this comparison. Wind speed is so site specific it is really important to get this bit right.

Once you have some idea of what the wind speed is in your area you can start to think about whether a turbine will work. Scale and height will very much be determined by what you are hoping to achieve and the size of community you are trying to power. There is a vast array of regional turbine manufacturers across the world particularly for the smaller 1kW-100kW scale machines. The larger machines, from about 500kW and up, are only made by about 10 large companies globally but may well be the choice for your site.

Power output from a wind turbine directly correlates to two factors: blade area swept and wind speed. As you increase blade size, potential output increases by the square of the diameter and as the wind speed increases the power available increases by the cube of the wind speed. The cube law of wind speeds means small differences in wind speed make a big difference in output.

A 20% drop in light intensity will produce a 20% drop in generated energy from a PV system; a 20% drop in wind speed, say from 5m/s to 4m/s will halve the generated output from a wind turbine. Simply put, the larger the turbine you build and the windier the site, the more power you will generate.

People often think about erecting a wind turbine above their town when it comes to generating community-owned renewable energy, and there are certainly many communities who have done this. In many areas, planning for wind turbines can be a complex and difficult process as opposition to their construction, mainly on visual grounds, can be strong. All the more reason to get the community involved.

Wind turbines might work to power your community, where you have appropriate wind resources and where you can find a site with a good location for a turbine. This might be a small machine that you construct yourself to charge batteries, or an array of large turbines that provides a significant chunk of energy for large town. Planning and implementing these very different scale projects will require very different inputs. It may be that for a small turbine project in a remote area, access to the technology itself and the finance to purchase it may be the biggest barrier. On larger projects grid connection and planning may be your challenges.

Planning a turbine project will begin with an analysis of the wind resource and the demand for electricity locally. From these two factors it will be possible to determine the optimum size for your turbine. If you have an estimate of the wind characteristics at your site, the free programme RetScreen will give you a first pass estimate of the generation and earnings for your system.

Things to look for in planning a wind turbine system:

- Grid connection – how much power can you connect to the local network? How much energy is required by the onsite consumer or host building?
- Access to land, away from buildings and trees to locate your turbine.
- Windspeed in your area.
- Grid independent systems will need some sort of battery storage and in some cases diesel or other backup.
- Are there any friendly farmers who could be a host site for the turbine?
- What will local people feel about it?
- Are there any environmental planning or zoning restrictions on the site?

HYDRO

The power of rivers and streams has been used since ancient times to carry out tasks that needed motive power, such as grinding flour or sawing wood. Driving much of the work during the early days of the industrial revolution, since the late 1800s it has been coupled to electric generation. It now provides the world's biggest renewable source of electrical generation – about 16.4% of global total electricity production in 2013.[19]

Much of this is provided by mega dams in countries like China and Brazil. Total global capacity is approximately 1,300GW with another 87GW under construction. Generation was estimated at 3,750TWh in 2013. Twenty-six percent of the installed capacity globally is in China, with Brazil hosting a further 8.6% and the US and Canada both having just under 8%.[20]

Currently there are four dams globally with an output in excess of 10GW, the biggest being the Three Gorges Dam in China with a peak power output of 22.5GW and generating 98.8 million MWh of energy in 2014. Given that a typical nuclear power station might have an output of 2GW, it is a pretty massive scheme. This eclipsed Itaipu Dam on the Paraná River bordering Brazil and Paraguay, previously having the highest annual output. At 14GW peak output it provides approximately 17% of the energy consumed in Brazil and 75% of consumption in Paraguay, producing 87.8 million MWh in 2014.[21]

Both these mega projects, whilst being an amazing feat of engineering, have their issues. Both required the relocation of many people who lived near the rivers; 1.3 million people were moved in the case of the Three Gorges Dam. Both sites also lost some unique features – buried under the waters of the Haipu Dam is the site of what was the world's largest flow waterfall, the Guaíra Falls. It was dynamited by the Brazilian government to enable safer navigation of the waterways. Understandably there were protests against the development. Under the Three Gorges Dam waters sit 1,300 archaeological sites.

Large dams do produce a lot of electricity, however there is concern that their negative impacts are effectively undoing their positive contribution. Sedimentation caused by the slowing of the river behind the dam means that nutrients don't make it downstream. Methane released due to the flooding of vegetation has the potential to contribute significantly to the greenhouse effect, given the scale of the lakes created. The river habitat is also badly disrupted, with challenges for fish migration and river ecology.

Hydro does come in a huge range of scales and many people around the world are accessing electricity not through these huge dams, but using much more human scale technologies. Picohydro and microhydro turbines are the smallest forms, and are used by thousands of people all over the world, particularly for rural electrification. Microhydro schemes don't need a large dam to make them work, and are generally geared to meet a local need rather than feed into the transmission system. There are already 85,000 small-scale hydro systems in China alone.[22] The pico turbines start at a few hundred watts of output, and go up to maybe 5kW, with micro turbines generating up to

---

[19]   *Renewables 2014: Global Status Report.* REN21.
[20]   *Renewables 2014: Global Status Report.* REN21.
[21]   www.itaipu.gov.br/en/energy/energy
[22]   Micro-hydro Power Factsheet – Practical Action, http://practicalaction.org/

100kW. In Nepal where half the population don't have access to the grid, electricity access is challenging purely because of mountainous geography. Microhydro has been a huge success and by 2012 there were 2,900 systems installed providing electricity to around 350,000 households. It has also helped to create around 80 local businesses able to provide all aspects for the construction and maintenance of the turbines.[23]

Which turbine you use on a particular site is very specific to that site. There are a range of turbine designs that cope with different conditions from high head low flow situations to low head high flow. Careful assessment of the water resource will be needed to determine the design and specification of the system and the potential output. To make a hydro turbine work you need a mixture of head i.e. vertical drop and water flow. So a mountain sited scheme may have a high head i.e. a few hundred metres drop between the water intake weir and the turbine. High head schemes often use 'pelton wheels' to extract the energy by directing the water flow through nozzles onto cups on the wheel. The wheel then drives a generator. For lower head sites, for instance sites where there was once a traditional water wheel, a crossflow turbine that works with a head of three metres may be more suitable.

There are turbines that can work in the run of a river also. The Archimedes screw turbine is one such design, which has become popular across Europe in recent years. It is a large piece of kit that sits in low weirs in the river and needs a large volume of water flowing down it to make it work. It is a useful turbine in that it has big spaces in the screw and it moves relatively slowly, which allows fish to pass through the turbine unharmed.

To get an accurate prediction of how much power a scheme will generate requires a full design study but a rough figure for judging the potential of a site is:

Power out (watts) = flow (litres/second) x head (metres) x 5

The head between where the water will be taken in and where it will flow out can be measured with a DIY laser level or an improvised water level with two transparent plastic bottles with their bottoms removed and fastened to a water filled hose.

The flow can be estimated by playing pooh sticks and timing the flow speed at the centre, estimating the cross sectional area of the river with a few soundings if possible and using the equation:

Flow rate (litres/second) =

23   Nepal Micro Hydropower Development Association, www.microhydro.org.np/index.php

centre flow rate (metres/second) x cross sectional area (metres²) x K²⁴

Where K ranges from about 450 for small turbulent streams to 750 for large slow ones.

This will only give you one day's figure and you will want to know what it is over the year.

If flow data is available from nearby sites a rough idea of how much flow we can expect from our site can be calculated by comparing the catchment areas of the two sites. Use the contours on the map to work out the watersheds. Estimate the areas by placing transparent graph paper over the map and counting squares. Alternatively Google Earth Pro has an area calculating function. If the two areas are not too dissimilar the flow pattern will scale with the area.

For low head sites, which need a high flow for a given output, often it is only sites with a pre-existing weir or dam such as old mill or lock sites that will be financially viable. The cost of constructing a fair-sized weir or dam to generate the head, and the difficulty of getting permission to create this sometimes make these sites unviable.

One thing that is common across all technologies and sites, is that if it is well designed a hydro turbine can give a steady output over the course of a year. Hydro can be deployed across a huge range of scales which means viable projects can range from a few hundred watts and up.

Factors in developing a hydro project:

- Expert help:
  There are many good guides on the subject of designing and installing small-scale hydro projects[25] and with some effort and a bit of engineering know-how the knowledge shared in them can easily be applied. However for many people it will probably be best to bring in an expert to make an initial analysis on your behalf.

- Resource analysis:
  Looking at what water resources you have in your local area will determine whether hydro is a potential for your community. If you have a river, looking at whether there were traditional mills in place on it will help guide you to places where it may be possible to reinstate them. It will be necessary to do an analysis of the water resource that you have, looking at flow rate and head, as well as calculations regarding

---

[24] Not all of the water in a stream river is flowing at the same speed. Therefore you need a factor to reduce the theoretical flow speed to get a more realistic output and that is 'k'. This gives the average speed of the water across the cross section as a percentage of the maximum speed. i.e 450 means the average speed is 45% of the maximum and 750 means the average speed is 75% of the maximum.

[25] http://store.cat.org.uk/product_reviews.php?products_id=420

the catchment area and the potential flow rates on the site. Seasonal variations also need consideration. From the calculations around water flow and head it will be possible to determine the scale and type of turbine that would be appropriate. At this stage you should be able to model the output of a potential turbine given the flow rates through the site, and work up an initial financial model to assess feasibility.

- Scale and design:
  With an expert's help a design will need drawing up for the particular situation and detailing any civil works that may be required, like creating weirs or diverting a section of the river into a race for the turbine. You may also need to construct fish ladders to allow the safe passage of fish past the turbine. Once you have a design you can then start to assess the costs of building the turbine. These cost assumptions will go into your financial model to see whether to progress the project further.

- Planning and other licences:
  In many areas extensive permits will be required to be able to construct a hydro project; these may include planning permission, an abstraction licence, a lease on the area of riverbed and as well as possibly discharge consents. Some of these take significant amounts of time to arrange and there is no guarantee they will be granted. It is not uncommon for the licensing process in the UK to take more than two years.

- Grid:
  Access to a grid connection will be required to make a hydro project work and the cost of this will need to be factored into the development. How much this costs and what it is possible to connect is also a determining factor in the project development. In areas where there isn't a grid to connect to, a scheme to deliver the power to the homes and businesses will also be required, with associated billing and metering set-up. Costs for developing a micro-grid and managing the business moving forwards will also need to be accounted for.

HEAT PUMPS

Most people in the industrialised nations already have a heat pump in their homes, but most don't realise it. It is in their kitchen and it's called a fridge. Heat pumps have been around for quite a long time being used commercially since the 1950s. By the end of 2013 there were nearly seven million heat pumps installed in buildings across Europe.[26]

---

[26]   Revolve 2nd Edition 2014 – European Heat Pump Association, www.ehpa.org

Heat pumps, as their name suggests, simply pump heat. In a fridge, you are trying to keep a big box and the food in it cool, so you pump the heat out of it and dissipate it into fins on the back of the unit that get hot. In the case of a ground source heat pump – a simple way to think of it is that the cold box is the ground around the building, into which you lay some pipes and the hot fins are the floor of your building where you pump the heat you extract from the ground. The temperature of the earth in northern climes stays at between 10-12°C all year round at a metre down, and the heat pump extracts a little bit of this heat – cooling it down. This passes through the heat pump, which then delivers that heat into the water circulating in the floors or radiators of the building. The water coming out of the ground loop may only be at 10°C, but it will be bumped up in the heat pump to between 30-40°C so that it warms the building up.

There is a range of ways to configure heat pumps, and there is a range of 'sources' to draw heat from. We can draw it from the air, the ground, a river or lake, or the sea depending on what is around us. Some heat pumps can be used two ways, to provide heat and to provide cooling. Heat pumps require electricity to run them, but well designed and installed, they will provide three to five units of heat for every unit of electricity that is put into them.

The low output temperature of many heat pumps means that the traditional heating equipment that has been installed alongside conventional boilers, like radiators, doesn't always work with them. Ideally a heat pump is coupled with technology like underfloor heating to give a warm and stable environment. A heat pump that has to pump heat up from 0-60°C will need twice as much electricity to do so as one that pumps heat up from 12-32°C. This is why air source heat pumps, although cheaper and causing a lot less disturbance to install, cost a lot more to run. In mid-winter, when you need them most, the air around the evaporator outside may be -5°C whereas the ground around the tubes will be 10°C or more.

To give you an idea of scale needed for horizontal ground pipes, for a well-insulated building you need an area for the pipes about four times the floor area of the building but there are a lot of factors that will vary this up or down. If there is insufficient area, it is possible to drill directly down or even drill several holes at an angle spreading out from near each other. If you are fortunate enough to have a suitable stream or river flowing past, the intake will only be about the size of a dustbin.

But heat pumps are not just used for heating buildings, they are also used in industrial application providing hot water for manufacturing processes, and extracting waste heat streams reusing the energy that would have otherwise been lost. They are also increasingly being plugged onto district heat networks, as well as being the sink for district cooling grids.

I love the ideas that are starting to be rolled out in some areas where heat pumps are being used to draw heat out of buildings that are too hot or need cooling, into a heat network, to be pumped back into buildings that need heat, and when there is an excess of heat on the network, it is being pumped into inter-seasonal storage like boreholes or building foundations.

That is exactly what is happening in Stockholm, Sweden, where there is already an extensive district heating network that reaches 60% of the city. Started in the 1990s, the networks are fed by a range of different technologies and fuels. Heat pumps form the baseload supply for the whole city with 420MW of installed plant, along with 200MW of bio fuel-fired plants. Oil-fired plants are used in times of high energy demand only. The source for the heat pumps is seawater, and large volumes are used to keep the temperature drop minimal.

They have also very successfully deployed the world's largest district cooling scheme. Alongside the networks of pipes that deliver heat is a network of pipes that deliver cooling. The heat pumps that drive this cooling network are drawing the cooling energy from the sea; the system also collects wasted energy from the heat network heat pumps, and has accumulators to store cold water over night, when cooling needs are less. By 2010 there were 600 buildings connected to the cooling network drawing about 330MW cooling load. The district cooling system has effectively replaced building mounted cooling systems (air conditioners) and in the process reduced the amount of energy used on cooling dramatically. If that cooling had been produced conventionally, it would have required five times more electric energy to create it. To top the whole thing off, an aquifer or borehole is used for storing heat or cooling for longer periods, and what is most amazing is that the whole scheme was built without public subsidy![27]

Sweden's large and highly developed schemes are an inspiration that are starting to be copied elsewhere. In London, one area is now being heated by a water sourced heat pump drawing heat from the River Thames. The first system of its kind in the Thames provides hot water for 150 homes and a large hotel in Richmond Park, by drawing water from two metres below the surface of the river.[28] The cool water drawn from the river is distributed around the homes and business by a network of pipes. Heat pumps close to the buildings extract a small amount of the heat from the river water which is then used to heat the buildings. The fact that cool water is circulated means the heat network required is cheaper and easier to build as the pipes don't need to be insulated and are therefore cheaper to install.

So heat pumps can be applicable to pretty much any building, and they are a very good place to 'start small but think big'. They have massive potential

[27] 'Showcase of District Cooling Systems in Europe – IEA Stockholm', http://www.iea-dhc.org/index.php?eID=tx_nawsecuredl&u=0&g=0&t=1438275073&hash=821e1a0a358f4969cf57439873cdfdf1e25b19d6&file=fileadmin/documents/DHC_CHP_Case_Studies/DC_Stockholm_City.pdf

[28] 'Heat from Rivers Could Supply Energy to Thousands of Homes', www.independent.co.uk/environment/green-living/heat-from-rivers-could-supply-energy-to-thousands-of-homes-9659501.html

across the world to join up heating and cooling needs and substantially reduce energy consumption. I think heat pumps could be a great place for a community energy company to start investing in heat technology.

What to consider when planning a heat pump project:

- Heat load – how much heat is needed?
- Energy efficiency – can you reduce the number above by being more efficient before you invest in the technology?
- Source – air, ground, water – what resources are available locally, is there enough space/access to the resource? In the case of water sourced, will you need a licence to extract the river or seawater?
- Tariffs: In some areas there will be grants or tariffs to drive the uptake of heat pump technology; these will help make the project viable.
- Electrical supply – your unit will consume electrical energy. Is there a suitable supply on site?
- If you are planning on delivering heat or cooling to more than one building then you will need to consider distribution and metering.

GEOTHERMAL ENERGY

Ground source heat pumps are sometimes wrongly referred to as geothermal heat pumps. Even vertical borehole heat pump systems get only a tiny fraction of their heat from the heat that wells up from deep underground. The heat that is extracted from the ground is almost entirely replenished by the sun from above. The world average for the heat from the sun hitting the ground is $186W/m^2$. The world average for the heat coming up from the ground is $0.087W/m^2$. This average, however, hides some areas where it is vastly higher. In parts of Iceland it is $0.35W/m^2$. True geothermal energy is used to generate electricity as well as most of the other uses that solar thermal energy is used for. In 2014 there was 12GW of installed capacity in the world that generated 61,600GWh of electricity. Another 12GW of capacity is being built.[29] Geothermal generated electricity has the special advantage among renewables of being nearly constantly available. The above figures represent a load factor of 59% compared with about 30% for wind and 20% for PV. In some countries, geothermal generation is a major contributor to national electrical generation: Iceland 27%, El Salvador 25%, Philippines 17%, Kenya 16% and New Zealand 13%.[30]

Although most geothermal electrical generating schemes are bigger than most community groups would want to tackle, some heating schemes are of a

---

[29] '2014 Annual U.S. & Global Geothermal Power Production Report', www.geo-energy.org/events/2014%20Annual%20US%20&%20Global%20Geothermal%20Power%20Production%20Report%20Final.pdf

[30] http://en.wikipedia.org/wiki/List_of_countries_by_electricity_production_from_renewable_source

modest scale. At a spa in Bansko, Macedonia, a scheme provides about 9MW of heating for a hotel, a swimming pool and a number of greenhouses.[31] At one time suitable sites were limited to those that had both hot rocks and naturally flowing ground water that made it possible to get the heat out. However modern drilling and fracturing techniques have made it possible to obtain heat from hot dry rocks, where there isn't also a source of flowing water. By injecting water into the hot drilled rocks it is possible to extract the heat available down there and pump it up for use on the surface. In the UK there are a number of developments like this taking place, one of which is in Cornwall at the pioneering Eden Project.

BIOGAS / ANAEROBIC DIGESTION

In China, biogas production began as early as the 1880s. By the 1950s the government began actively promoting it and today they are actually the world leader in biogas production. It amazes me that over 40 million residential biogas fermenters have been built in China to date. Tens of millions of families use farm and household waste to make clean cooking fuel in these backyard fermenters. Yet only 19% of the potential for biogas had been utilised as of 2010.[32] Germany generates as much electricity as two nuclear power plants with the gas produced by decaying plant matter and animal slurry. Biogas is an old technique applied around the world in low-tech and high-tech ways to deal with a range of materials that would otherwise be considered as waste problems. However, in true alchemist fashion, it turns muck into gold, producing a couple of really useful outputs in tackling the waste stream: gas and fertiliser.

When you put organic matter that is high in nutrients into a sealed container, with the right microbes, and keep it at roughly body temperature, it will begin to ferment. This process, called 'anaerobic digestion', is a natural process, actually going on in our bodies, and it produces a gas output – bio methane. Methane is a greenhouse gas 30 times more potent that carbon dioxide and is also a very useful fuel when burned. Collecting it and burning it, to make heat, electricity, motion or a combination of them all, is a great way to prevent it being released. It is also a great way to extract a useful resource from materials which would release their carbon anyway as they rotted down in the air.

Germany has the most advanced biogas market in Europe, producing over 50% of the total biogas produced in Europe from nearly 8,000 digesters.[33]

---

[31] *Geothermal Energy: Utilization and Technology.* Eds. M H Dickson & M Fanelli. 2003, UNESCO.

[32] Chen, Yu, Feng, Yongzhong, Sweeney, Sandra and Yang, Gaihe (2009). 'Household Biogas Use in Rural China: A Study of Opportunities and Constraints'. In *Renewable and Sustainable Energy Reviews* Vol. 14, Issue 1, January 2010, pp.545-549.

[33] *Biogas: An All-rounder – New Opportunities for Farming, Industry and the Environment.* 3rd fully revised edition. 2013, RENI.

In 2012 these plants supplied about 6.5 million homes in Germany with electricity and created over 40,000 local jobs.

Lots of the biogas plants are sited on farms where slurries and wastes, that could otherwise be causing environmental pollution problems, are recycled into gas. In some cases specific crops are grown to add to the mix of material put into the digester. Some digesters are fed sewage wastes, whilst others are fed municipal wastes, like household compost. The stream of material and the quantities will have a huge impact on the quality of the gas that comes out, as well as the size of the plant and the volume of gas produced. The process is a living process, which if not tended and fed new material will stop working. As such it needs active management and someone to look after it. That's one of the reasons why they can work so well on farms, because someone is there tending animals and removing the wastes every day anyway.

The produced gas can then take a number of routes to becoming useful energy. It is sometimes burned on site to produce heat and electricity in a CHP plant. The heat is provided to local homes or businesses via a district heating network, and the electricity is either supplied directly to nearby users or straight to the electrical grid. In other cases the gas is piped elsewhere for combustion in a CHP plant, or cleaned and injected into the natural gas grid. It needs to be cleaned before sending into the gas grid as it is only 50-60% methane, the rest being made up of carbon dioxide as well as small amounts of nitrogen, oxygen, water vapour and even smaller amounts of ammonia and hydrogen sulphide. These impurities make it very corrosive if not treated. In Germany, around 100 plants feed 67,000 standard cubic metres of biomethane into the grid every hour. Once in the grid it can be used for many different uses including transport. In Sweden the public transport in 10 towns runs on biomethane. One hectare of maize turned into biomethane allows a vehicle to travel around 70,000 kilometres. That's the equivalent of one and a half times around the world!

In China and many other countries – Nepal, Bangladesh, India, Sri Lanka to name a few – smaller much more low-tech systems are used. These are backyard digesters that take all of a homestead or community's waste stream, human and animal manure as well as agricultural residues. Exactly the same principles: the wastes are added to a sealed chamber where they ferment to produce gas which is then drawn off for lighting and cooking. Sometimes they are 'three in one' systems combining a toilet and a pig pen with the digester. Access to energy in rural areas is often limited to biomass and the produced gas provides valuable energy for cooking, preventing woodland being cut down for fuel. The digesters kill many pathogens during the digestion process so they can help to solve human health issues at the same time as producing useful fertiliser for the crops. The gas is also often used for lighting and sometimes to drive machinery. The digesters are built using local materials, skills and resources. What could be more sustainable?

At the other end of the spectrum, modern biogas production facilities are often technically advanced plants, which combine a mixture of wastes with energy crops such as maize. Using computers to control the process, adding precise amounts of the various feedstocks with automated loaders to get the mix just right, enables a very high production of useful gas.

In Germany there were around 130 communities that had set up collectively owned biogas facilities by 2012. In 2008, the villages of Wollbrandshausen and Krebeck decided to jointly construct a biogas plant. So residents established a co-operative with 251 members and jointly constructed a biogas plant with a capacity of 1.8MW and a cost of €12 million. The power generated is fed into the public grid and the heat is used locally.

The scheme brings people together in the two villages and means that local farmers work together to supply the plant. On top of this local people are employed to run the plant. To keep it working, eight farmers supply slurry, and 30 farmers grow and supply the 30,000 tons of maize that are required to run the system. The heat is delivered by a heat network running 10km between the villages and serving 260 customers. By summer 2012 the plant had also produced over 30 million kWh of electricity.

There are multiple local benefits of establishing a biogas production facility in your area. Turn waste into fertiliser and energy, and potentially create employment at the same time. Done right it could insulate your community from the global energy market, keeping the money that would have otherwise have been hoovered up by large corporate energy companies circulating locally.

So whether you are rural or urban, in an industrialised nation or not, biogas is likely to be applicable in your community. However it might not be the easiest project to develop, build and maintain. Unless considering the simplest forms of digesters, guides to and plans of which are readily available on the internet,[34] specialist help will almost certainly be required in the planning and design of any biogas production facility. These projects are large and complex to develop, requiring many variables to be explored and clarified before a project is ready for financing.

Where do you start when considering a biogas project in your community? Is there a ready supply of organic matter that can feed the plant?

A wider range of sources can be used, from kitchen and food waste, sewage slurry, agricultural slurry, agricultural by-products as well as energy crops. The success of the plant will depend on getting the right mix of materials in the right quantities into it. Considering the amounts of waste that are produced particularly by cities all over the world, and how much of that is organic matter, this ought not to be a problem. In fact in many countries where organised waste facilities simply do not exist it is almost certainly a huge opportunity.

---

[34]   Practical Action – Biogas, http://practicalaction.org/biogas-fuel

Do specific energy crops need to be grown?

To get high production of gas from the plant it may well need a specific energy crop to be grown to be mixed with the waste streams. If it does then a contract to grow the crop will need to be put in place with local farmers to ensure the supply in the years ahead. The viability of the scheme will rely on the robustness of the contract, so it will need careful consideration.

Are there any local partners you need to develop your project?

Local farmers and the local municipality will both be needed in some form or other to create a successful biogas scheme. Consider the possible inputs, sites to host the plant, permissions, possible off takers for the power and heat produced. Planning permission will be required, and the extra vehicle movements need to be accounted for.

Does your municipality already collect and separate organic wastes that could be used for the plant? If yes, where do they currently go? If not could that be achieved in a cost-effective way to feed the plant?

Municipalities will generally hold figures on the wastes that they handle on our behalf, and may already be considering such a scheme. Their input and co-operation will be essential in making any scheme involving household kitchen waste successful.

Scale and size

The volume of material that is available for the plant will determine the size and scale. It will also be necessary to consider the grid connection and the amount of power you could produce and export to the grid or provide to local users. If the scheme will also produce heat, are there enough local heat consumers to make it viable? Will a new district heating scheme be needed to deliver the heat? If yes this will need costing and designing. The local community will need to be convinced of the benefits of the new plant and of the viability of getting heat from it. If you were planning to inject the gas into the gas network, a suitable connection point will be needed.

Construction and finance

Experts will be needed to construct the plant, and finance will be needed first to carry out the risk development works, in advance of actually constructing the project.

Maintenance and running

The plant will need daily care to ensure it stays running optimally and safely. It will need regular deliveries of materials to be digested and regular removal of fertilisers and slurries produced. It is likely that the plant will create some jobs in the process. Billing of heat, and gas or electricity, will also need to be factored into the process of running the plant.

BIOMASS

Using combustible material from plants is still the most widespread use of energy on the planet, with many people around the world still relying on biomass for cooking and heating. In industrialised nations the use of biomass was displaced by coal and other fuels. However in some areas biomass use has been reinvented and reapplied for modern homes and lifestyles with spectacular results.

Austria has been one of the areas pioneering modern biomass production and use. By 2011 in the state of Upper Austria there were 44,000 automatic biomass boilers in operation in residential, commercial, and public sector buildings. Half of these are fuelled with pellets, and half with wood chips. They have a total installed capacity of 2,100MW and produce 3,400,000MWh each year. A whopping 47% of total heating demand. The use of biomass that is produced locally means that the area avoided fossil fuel costs to the tune of around €1 million, consuming 1,150,000 tons of biomass fuel, and achieving CO_2 emission reduction of 1 million tons per year.[35]

Upper Austria has had the right set of circumstances to drive this development: forward thinking and entrepreneurial forest owners and farmers; strict emissions and efficiency standards; and stable policy support to help them achieve critical mass. The region also has a good level of forest cover with 41% covered in woodland, compared to the US with 30% cover, or Sweden's 67% cover and Germany with 30% cover. Interestingly the business model that emerged to deliver many of the biomass projects in the region of Upper Austria was to use co-operatives. Groups of farmers and forest owners coming together to develop, build and operate small-scale biomass district energy systems. Providing heat to their local communities – local government buildings, schools, businesses and housing. Today there are more than 310 biomass district heating networks in operation.

Significant advances have been made in the use of biomass fuel from the open fire to fully automated wood heating appliances, much of it in Upper Austria. There are a range of scales, technologies and fuels that can be applied depending on circumstance. The great thing about modern biomass systems is that they can be entirely automated, and are much more efficient than traditional combustion systems, with more than 80% of the energy put in being utilised for the end goal.

---

[35]   Biomass Heating in Upper Austria Green Energy, Green Jobs – OO Energiesparverband, Linz, Austria, www.esv.or.at

Fuel types

There are a range of fuels that can be easily used in a biomass system, such as wood or other agricultural products like straw. Your location will determine what is available and might therefore be viable.

Wood

Trees are a very stable form of agriculture and planting species that can be coppiced or regularly harvested can become a very good source of fuel wood. Generally fuel wood comes in three forms:

- Logs
- Chips
- Pellets

The less the fuel has to be processed the cheaper it will be to purchase, however it will also be harder to store and handle. Logs for example are bulky and may require manual handling to load them into a boiler, whereas woodchip can be moved much more easily using machinery and automation. Logs may be a choice for a community energy project in your area if there is access to local woods that need managing. The system will need lots of human input to process the logs and keep it running. Storage of the logs will also be an important factor, as logs easily absorb moisture and moist fuel does not burn well. Woodchip can be made from a wider range of wood than logs meaning more of the tree can be utilised. It will be cheaper and can be moved and ignited more easily, making it an easier fuel to use. However storage and moisture content will be a key factor in making it work well. Pellet fuel is the most stable and refined form of biomass fuel, generally made from compressed sawdust and produced and inspected to very tight standards. It is the most expensive form of fuel and unlikely to be made in your community although there are over a hundred manufacturers in Europe. It is however the most hassle free fuel in terms of operation. The uniformity of the pellets means that the boilers can be turned down to 30% without going out or losing efficiency, and boilers can be designed that condense the water vapour in the flue gas adding about 15% to the efficiency. There are specialist boilers designed to handle the various different types of fuel.

Agricultural wastes and energy crops

Agricultural by-products like wheat straw can be used to fuel a boiler, and can be a good use of a material that would otherwise simply rot. Specialist boilers are available specifically designed to burn bales in various sizes. Specific short growing crops like the African grass miscanthus or willow coppice can be cultivated to provide fuel for boilers. Often they require specialist machinery to process them for drying, storage and burning. At the Institute of Biological,

Environmental and Rural Sciences (IBERS) in Aberystwyth, Wales, they have made pellets out of miscanthus grass with a standard wood pellet machine which burn on a standard pellet boiler.

Technology

There are a range of systems that can be used to burn biomass, from manually loaded batch heaters that need daily attention, to fully automated systems that only need to be fuelled once a year. Generally both wood pellet and wood chip systems will feature automatic ignition systems. This means that the boilers can be started and stopped using a programmer, making them very user friendly, and the transition from fossil fuels much easier. Biomass can be effective when installed in a range of different scales and applications:

- Individual homes – systems with 5 kW-25kW output, using wood pellet
- Farm systems – with 15kW-50kW output, using logs
- Public and commercial buildings – with 20kW-300kW output, using woodchip
- District heating schemes, 100kW-MW output, using woodchip

Fuel storage and handling

Depending on the type of fuel handling, storage will be very different. Woodchip is often stored in barns or silos, and needs to be the right moisture content when being stored. Often augers (worm screw) will be used to deliver the fuel to the boiler, collecting the chip from a central point. This is similar with pellets, but as they are even more predictable there are an increasing range of ways to store and transport them. In its most advanced, wood pellet is moved from large external pellet storage tanks remote to the boiler. This is achieved using vacuum technology, sucking the pellets from a flexible fabric storage tank directly to the boiler. Auger systems are also used for delivering wood pellets where the runs are shorter.

What to think about in planning a biomass system in your community:

- Loads:
  What needs to be heated? A survey of the building you are looking to provide heat to will need to be undertaken to establish exactly how much heat it needs. If there is more than one, then repeat the process and also consider heat delivery. Could any energy efficiency be carried out first? This should be looked into as if applied second it may mean that you have invested in a plant that is oversized for the task once the efficiency work has been undertaken.

- Fuel sources and supply contract:
  Which fuel is optimum for our system will depend on the scale of the

system, the degree of automation required, and what is available locally. Locking in a supply contract for the fuel selected for the project will be crucial. Ensuring a supply of fuel at the right quality and price will help to make the project predictable in terms of costs and outputs.

- Design and installation:
  When the loads are known it will be possible to get a design drawn up for the system. Specialists will be needed to specify, design and construct the plant.

- Operation and Maintenance (O&M):
  Once you have designed and constructed the system it will need looking after. Maintenance will be needed on pumps and heating appliances. The system will require administrating, billing the users for the heat they consume, making sure the fuel is delivered and the plant is running well. Often the company who constructed the unit will be bought in to provide the O&M service.

COMBINED HEAT AND POWER (CHP)

Combined heat and power does exactly what is says on the tin. It is a device with combined outputs of heat and power. Conventional power stations making electricity often have a waste heat 'problem' and as such need huge cooling towers, which release that heat into the atmosphere. Hence the efficiency of most large power stations will struggle to reach 40%.

CHP on the other hand sees that heat as an opportunity. Instead of being vented into the atmosphere, it is used to provide useful heat in other areas. CHP power stations come in a range of sizes from micro systems to large industrial units. They are generally sited near to where the heat is needed, i.e. near to buildings, homes and factories. Consequentially a CHP unit can reach 75% efficiency. For every 100 units of energy that is put into the system, 30 units will be turned into electricity, and 45 units will be turned into useable heat. This is quite a step forward on a traditional power generation system.

They can be powered with a range of fuel sources, biomass, biogas, or fossil fuels. The fuel is burnt to create heat, which is used to drive a turbine or reciprocating engine to create electricity. The 'waste' heat created is then dumped into a district heat network or other heat load, like buildings and swimming pools that are attached, putting what would traditionally be lost to good use. At the smaller end of the scale, up to 1.5MW of electrical output, CHP units come in packaged units that can be bought off the shelf. For larger projects involving steam or gas turbines they will generally be bespoke and add considerably to the upfront design work.

A lot of work has been put into designing small CHP units that use hydrogen fuel cells to generate the electricity. The methane gas fuel is first reacted with water to produce hydrogen that is then fed to the fuel cell. The hope is that the overall efficiency will be the same but the division between heat and electricity will be nearly equal. This would be a better match for domestic users who in summer have only a small heat load. The units will, in addition, be smaller and quieter.

Designing and installing a CHP unit in your community would start with an assessment of the heat load. A steady heat demand year round will be needed to make a CHP system work efficiently, so good analysis of the demand is essential. This will involve looking at the heat demand of the different energy users in the area, along with when they are needing the heat. From this we can calculate a baseload heat demand and a peak load heat demand that enables the sizing of the plant. Once the base heat load is calculated then it will be possible to determine the electrical output. Different technologies have different heat to electrical ratios depending on the scale of the project and the type of fuel used. The simplest fuels to use are gas and oil, but they may not achieve your CO_2 goals or provide you with a local fuel source unless using biogas. The simplest fuels are of course the most price volatile, and the most expensive potentially. Solid fuels such as biomass will need handling and delivery and contracts in place to guarantee security of supply. Once you have sizing calculated, you can look at the costs of the various elements; engine or turbine, electrical generator and heat recovery units, and whether it stacks up as a project for investors. A designed CHP unit can achieve paybacks as fast as four years, so they can be a very attractive investment.

If the CHP unit is not just connected to one large heat user, it may need to be coupled to a district heating system, which will also need sizing, costing and designing. It will also need a grid connection capable of taking the electrical output of the system, so a conversation with the local network operator will be required.

A good place to start in your community is to look for the highest heat users. Starting with them will mean that the project is more likely to stack up and achieve the baseload demand required to make the system function quicker. If they have new boilers in place, long-term energy supply contracts or short building leases in place, it may be challenging to make the project work.

It is likely that you will have to work with multiple stakeholders across your community to make a CHP based project work, and of course the more parties you have around the table the more complex it is to get a deal in place.

DISTRICT HEATING

District heating schemes deliver heat (or cooling) to multiple users, be that homes, offices, or factories. I think that this will be a key technology in making our communities more resilient in energy, financial and climate terms, even though they are energy distribution systems not renewable generation. Often the homes or buildings are connected to a single large heat source, or multiple larger heat sources. In many cities around the world buildings are connected to a gas network and have their own heat appliance – a boiler – in the building. In a district heating network, heat is delivered directly to the building through insulated pipes which run under the ground instead of the gas pipes. A larger boiler or other heat source is connected at the other end. The network of insulated pipes connects every building, and heat is delivered to the building through a heat meter. The heat meter measures the temperatures of the water flowing into and out of the building as well as the speed of flow. From that it is possible to determine the amount of heat being consumed.

Installing a district heat system can be an expensive process when it is being applied to existing buildings, as roads have to be dug up to install the distribution pipework and heat meters have to be installed on each building receiving the heat. On a new development it would not necessarily add extra costs in that services would have to be installed anyway. The cost of the insulated pipe is one of the major costs of a district heating plant. Therefore plants that distribute cold water, and use localised heat pumps, reduce the cost of installing the network, as the pipes are cheaper and cost less to install.

District heating will be crucial in developing low-carbon locally fuelled heating systems for our neighbourhoods. Denmark's impressive transition was in some way underpinned by district heating. Over the course of 30 years they went from none to 60% of homes heated by district heating schemes. One small Danish town, Marstal, has been a pioneer in combining inputs into its district heating system. They have $33,000m^2$ of solar thermal, a 20MW biomass plant and a heat pump feeding into the network. They also have two large-scale heat storage areas, large insulated pits that store the excess heat produced in summer in water or sand. They achieve 100% renewable heating for over 95% of the towns 1,650 dwellings, with over 50% of the energy coming from the solar panels.

One of the joys of district heating is that a range of heat sources can be used, and heat storage can be added to cope with excess heat across the network. Combined with a district cooling network as there is in Stockholm, the system can extract waste heat and provide that heat where it is needed. Very neat.

Brilliant systems once complete, however, planning a heat network is a complex process, requiring considerable upfront design and analysis.

Support from the people

In order to make a project like this happen you will need to build support in the community which will be affected by the scheme. In many places people will never have heard of district heating and will be used to dealing with their own heating appliances and systems. The proposal of a collectively owned heating system will take some time for people to understand and accept.

Presenting the information on the ideas in a clear and accessible way will be essential. Showing examples of other communities that have already benefitted will really help the case. In some areas legislation enables new district heating schemes to drag residents into the scheme if over a certain percentage, say 90%, are signed up. In many other cases it will not, so a campaign to bring people along will be needed.

Loads

What needs to be heated or cooled? A survey of all of the buildings you are looking to provide heat to will need to be undertaken to establish exactly how much heat each one needs. Could any energy efficiency be carried out first?

Heat sources

Whether you go for a straight boiler or a combined heat and power system will very much depend on the fuel, location, grid connection and type of loads. However in all cases there should be consideration of what heat sources can be used for the network. As we have heard, heat pumps, biomass plants, biogas plants, solar thermal as well as fossil fuel powered heat generation can all be fed into the system. What fuels and heat sources are available locally will determine what is possible. So if you are near the sea or a river that may be an option; if you are in a rural community you may have biogas, biomass and solar thermal.

Design and installation

Design and construction of the system is a specialist task, and one that will need an experienced contractor – or a number of them. It has the potential to be a disruptive process with streets needing to be dug up for the system to be installed, and new meter cabinets and heat exchangers will need to be integrated into the buildings' existing systems. Areas that don't have wholly paved streets will be cheaper to integrate these systems into, as digging a trench in a field is a lot cheaper than digging the same trench in a city street.

Operation and maintenance

Once you have designed and constructed the system it will need looking after. Maintenance will be needed on pumps and heating appliances, as well as heat transfer units. The system will require administering, billing the users for the heat they consume, making sure the primary energy source is delivered and the plant is running well.

ENERGY STORAGE

Storage of energy will be crucial to balance the fluctuating generation created by renewable energy. It is not something many communities will be able to address, however as our success grows it will become more important.

Energy storage can be looked at in two forms: electrical storage and heat storage. I address each briefly below.

Heat can be stored in a range of simple ways. Many of us already have heat storage in our homes in the form of a hot water cylinder that stores the domestic hot water. This model can be scaled up for use by communities in the form of seasonal storage. A giant-sized hot water cylinder or 'thermal store' is often used for this role, the taller the better – as this allows for good stratification – and more useful heat off the top. It can provide the balancing load for a range of heat inputs and potentially provide heat for many dwellings.

Heat storage can also be achieved with 'pit storage' – large covered ponds with huge amounts of water in them. The ponds are insulated on the top to avoid heat loss and store large amounts of heat when excess is being produced in the summer months. In Marstal, Denmark, the last addition to their heat storage portfolio contains 10,000m³ of water, and stores a significant amount of summer heat to provide for winter use for a small town.

Borehole storage is another idea, where a series of boreholes are drilled and the heated water is piped down into the ground via a network of pipes. The heated earth then acts as the storage medium, with heat being extracted in winter as is the case with Drakes Landing in Canada. A variation on this is aquifer storage, where instead of using the ground as the heat store, an underground aquifer is used instead. Two boreholes are drilled into the aquifer then heated water is injected into one of the boreholes to charge the store and cold water is extracted from the other. This can be a very effective method, however if the water in the aquifer starts to flow, it can strip away the stored heat.

Generally the costs decrease as you go from tanks, to pits, and then to boreholes. However the density of the heat stored also reduces, so you need a larger volume to store the same amount of heat energy.

Considerably more heat can be stored in a given volume by incorporating a 'phase change material', essentially a material that melts at a specified temperature. A purified form of candle wax that melts over a narrow range of temperatures close to the temperature you want to store your heat can be used. Over the 5°C range 28-33°C the material Rubytherm SP31[36] stores 11 times as much heat as water.

Storing electricity is commonly achieved in batteries, and on an industrial scale in pumped water. There is pumped storage all over the world, being used for grid stabilisation and balancing already. By 2012, there was around 130GW of pumped storage capacity globally.[37] As a technology, it is very simple, requiring two reservoirs with a significant height difference between them. In between a turbine is used to pump the water uphill in times of

---

[36]   www.rubitherm.de/english
[37]   International Energy Statistics, www.eia.gov/cfapps/ipdbproject/iedindex3.
       cfm?tid=2&pid=82&aid=7&cid=regions&syid=2004&eyid=2012&unit=MK

excess electrical production. This pumped water is then released back down to generate electricity when demand is high, and it can be turned on in seconds. It can achieve efficiencies of over 80%. However this design has the limitation of needing a mountainside to make it work, so will not be applicable in many locations. A new 50MW scheme has been approved in Wales in a disused slate quarry[38] showing that such schemes do not need to be of enormous size.

Batteries are a part of our everyday lives; all of the handheld devices we take for granted have batteries embedded into them. Increasingly large-scale batteries are starting to be used to support the electrical network, delivering power quickly at peak consumption moments, and storing excess energy when there is surplus. Europe's first and largest commercial battery plant has been launched in Germany; it stores and can return 5MW of power for one hour.[39] In Chile there are already a number of large-scale battery backup plants used to stabilise the grid; able to dispatch 20MW of power for 15 minutes to adjust for failures on the network or surges in demand.[40] In Sendai, Japan, a lithium-ion battery store will open in 2015 with a maximum power of 40MW, and it will be able to run at that rate for half an hour. In Texas the Notrees Battery Storage Project uses lead-acid batteries to deliver 36MW of power for 40 minutes at a time.[41] These are significant chunks of power, and if enough of them are deployed across a network, they will reduce the demand for the most expensive peak demand generators, and their associated pollution that is needed to get the required second by second balance between demand and supply. The short time they can run at full power is adequate for this as it allows time for a conventional plant to start up. However they are not a solution to the week or longer variations in solar and wind generation.

Most of the large-scale systems deployed use traditional battery technologies, like lead acid or more modern versions like lithium-ion technology. These cells are limited by their construction and volume, and to move beyond this, flow batteries have been developed. Flow batteries store the energy in the liquid part (electrolyte) of the cell rather than the solid bit (anode or cathode) as is the case with traditional battery designs. This means that the technology becomes easily more scalable – add more electrolyte to add more capacity. The technology has already been deployed in some commercial situations, like in a green onion processing factory – Gilles Onions! They have used a combination of technologies to turn their waste products into electricity, and use a flow

[38] 'Green Light for Welsh Hydroelectric Power Station', www.ft.com/cms/s/0/e210981a-1641-11e3-a57d-00144feabdc0.html#axzz3OTEyrIT

[39] 'Germany Opens First Renewable Energy Storage Facility', www.rt.com/news/188372-germany-energy-renewable-batteryplant/

[40] 'Bringing Stability to the Grid: AES Energy Storage Angamos Battery Energy Storage System', www.energystorage.org/energy-storage/case-studies/bringing-stability-grid-aes-energy-storage-angamos-battery-energy

[41] Smooth Operators, www.economist.com/news/technology-quarterly/21635331-matching-output-demand-hard-wind-and-solar-power-answer-store

battery to shift their demand peak and thus reduce their bills dramatically. The flow battery makes 600kW of power available at any time during the day.[42]

Electricity can also be converted into heat, or more precisely heat differential. The heat storage of electricity using heat pump/engines and heat stores is being pioneered by a number of companies. The system developed by Isentropic Ltd uses the following process: During the charging cycle, the heat engine cools one large cylinder to -120°C whilst at the same time heating a second cylinder to +500°C. The two cylinders are full of crushed gravel, and when the electricity is needed the process simply reverses, with the heat differential driving the heat engine to produce electricity. By making all parts of the system as reversible as possible, round trip efficiencies of 90% can be achieved. Using gravel as the heat store keeps the system cheap to construct and easy to expand to store more.[43]

Another method of storing energy is compressing air into an underground cavity such as a washed out salt dome as is done at Huntorf in Germany[44] and MacIntosh in Alabama USA.[45] Both these systems use the compressed air mixed with natural gas to feed a gas turbine to generate electricity. Nine other proposed schemes are in various stages of progress.

Energy can be stored by liquefying air. Off peak electricity is used to liquefy air using the Claude compression cycle used to produce millions of tonnes a year of oxygen for the steel industry by distilling liquid air. It is stored in large vacuum insulated vessels such as is used to transport liquefied natural gas. When it is required to recover the energy the liquid is compressed to a high pressure. Because it is a liquid this does not take much energy. It is then heated to produce high pressure air to turn a gas turbine which drives a generator. If electrical heat is used the round trip efficiency would not be much more than 50% but when the heating can be done with very low grade waste heat then this improves to 70-80%. A 300kW, 2.5MWh pilot plant has been in operation since 2012.[46] An engine that runs on liquid air has been designed by Peter Dearman and it is hoped a prototype vehicle will be road tested in 2015.[47] Its only exhaust is air. Because a huge amount of energy can be stored in the form of liquid air cheaply and in a small space almost anywhere it will be possible to capture wind and solar energy when it is available for use in vehicles months later.

---

[42] 'Peak Shaving and Demand Change Avoidance: Prudent Energy Vanadium Redox Battery Energy Storage System', www.energystorage.org/energy-storage/case-studies/peak-shaving-and-demand-charge-avoidance-prudent-energy-vanadium-redox

[43] Isentropic Ltd, www.isentropic.co.uk

[44] 'Compressed Air Energy Storage to Buffer Wind Power', www.bine.info/en/newsoverview/news/druckluftspeicher-sollen-windstrom-speichern/

[45] CAES, www.powersouth.com/mcintosh_power_plant/compressed_air_energy

[46] 'GE Partners With Highview for Liquid Air Energy Storage', www.greentechmedia.com/articles/read/ge-partners-with-highview-for-liquid-air-energy-storage

[47] www.media.wix.com/ugd/96e3a4_0d7a380f317b4768a7572e5a0dac5fb1.pdf

Turning electricity into gas is another way to store the energy. Creating 'synthetic methane' or hydrogen which can be used directly for other energy needs is a good way to do something useful with excess generation. In many nations we already have extensive gas networks which can act as the storage mediums for the gas injected. In Frankfurt, Germany, there is already one commercial plant that injects hydrogen into the natural gas grid. When there is excess production on the grid, the plant uses electrolysis to create the hydrogen.[48]

For me perhaps the most exciting opportunity is the potential to use a fleet of electric cars to back up our homes and smooth peak demand. Electric car pioneer Tesla Motors has this goal firmly in their sights. They have just started construction of a new $5 billion 'Gigafactory' to produce lithium-ion batteries. Once complete it will produce more batteries than all of the other production facilities globally combined. This massive scaling of the technology is hoped to dramatically drop the price of the batteries. This in turn will make their electric cars more affordable, as well enabling home energy storage.[49] Affiliated company Solarcity has already started to offer Tesla battery backup systems to commercial customers to help them reduce grid import at expensive times of day. At some point in the not too distant future 'grid defection' may become a reality – where people decide to minimise their use of the grid by installing a home energy storage system. Solar plus battery and grid will most likely change the face of the energy markets in the US as well as Europe.

ENERGY EFFICIENCY

Energy efficiency is the sleeping giant. Using less energy makes all of the above easier. It is absolutely fundamental to our success as communities that we stop the energy waste in our homes and businesses. Put simply our leaking buildings throw energy and money away, and the effect of our local waste always has to be amplified up the supply chain. It has an impact many times larger when we consider the giant energy machine feeding us our heat, light and power.

The cheapest form of energy is not to use it in the first place – 'the negawatt is cheaper than the megawatt', as Amory Lovins puts it. Much has been written about turning our thermometers or aircon down and switching our light bulbs, which are all important actions. I can't help but feeling, however, this is one area that needs a real push from top down as well as actions from the ground up. Take the example of Japan (see page 107) – huge reductions in energy consumption and peak load were achieved with effectively no investment – led by the Prime Minister as a national priority. Surely governments around the world should hold energy efficiency as a national priority and give it appropriate frameworks to make it work and help people adopt it en masse and at speed.

---

[48]   THÜGA Power-to-gas Plant, www.itm-power.com/project/thuga-power-to-gas
[49]   'Tesla Closes on Free Nevada Land for Gigafactory', http://fortune.com/2014/10/28/tesla-closes-on-free-nevada-land-for-gigafactory/

The multiple benefits in terms of jobs, fuel poverty, balance of payments and positive health outcomes, would surely outweigh the costs involved.

There are schemes out there launched with these aims in mind, however it seems that in the UK they have fallen well short. The 'Green Deal' is a scheme designed to deliver appropriate energy efficiency measures to homeowners. Following an assessment of the building, homeowners are recommended a package of measures that bring down their energy consumption, and are offered funding to implement them. The funding is a loan scheme linked to the property, which is paid back through utility bills over a number of years. All this sounds like an ideal package, and in theory it is, however the interest rates involved in the loan mean half the capital repaid over the life of the loan is paid in interest payments. If the government had fixed a low interest rate on the scheme it may well have been a huge success.

Surely we could run schemes like these in our communities? Working in our own towns to identify areas that are in need of complete home retrofits, identifying the measures needed in each property, and aggregating them into a portfolio. Raising money from the community to deliver the measures, and implementing them at no cost or partial upfront cost to the homeowners. Giving a return to investors and returning capital by collecting a monthly payment from the householders. The finance should be set so that the monthly payment is less than or equal to the savings that are being made in having the measures in place.

The reason I put this last is because it is the most complex business model and group of technologies to take from nothing to a completed project, and I don't know of a community that has yet pulled this off. However this is surely where we must all set our sights.

What can be done?
It is possible to dramatically reduce the energy consumption, energy costs and CO_2 emissions of our homes and buildings with simple technologies that are available today. A huge amount can be achieved with fairly modest investments. It is not uncommon to drop the energy requirements of older buildings by over 50% with a holistic energy efficiency programme, and it is possible that this could be as high as an 80% reduction in energy requirements. An energy efficiency programme will look at the building in detail, its construction, the materials in use, windows, doors etc. and appliances. Taking measurements and doing the sums it is possible to determine the exact places where measures need to be applied to prevent heat loss from a building, and increase the 'U' value or insulating value of the walls and other surfaces.

Building fabric
25% of heat is lost through un-insulated roof spaces, 35% through un-insulated walls, 10% through poor windows, 15% through doors. A simple place to start.

Heating systems
Older heating systems can be greatly improved upon, both in terms of the boiler and the heat delivery system. Condensing boilers extract more heat from the fuel by using larger heat exchangers. Boiler and radiator controls can dramatically reduce heat use in areas of the building you are not using.

Appliances
When purchasing new appliances choose the ones that use less energy. Replace standard light bulbs with compact fluorescents or even LEDs, as this can lead to a massive reduction in the costs of energy.

All of these elements could be offered as part of a loan-backed scheme for homeowners who couldn't afford to invest in the measures themselves. I look forward to seeing the first communities as they roll this out.

CHAPTER 29

BUSINESS AND FINANCIAL PLANNING

You have pulled a team together, found a site and identified which technology might be possible. Now it is time to work out whether the business model will actually be viable. To do that you will need plenty of information about your possible projects and will almost certainly have to make a whole range of assumptions too. This section is based primarily on setting up a local generation project, but many of the aspects will be the same for other types of business.

A local energy co-operative or company is a business, with all the usual challenges and requirements that come with it. From cashflow to tax, from contracts to marketing, your new business will have to cover a lot of bases in order to function. Before we get to that point, you need to do the sums and work out whether you actually have a business!

Business planning is the process of doing that: defining exactly what you are proposing to do, what it will cost to achieve, who will do it, what income it will generate and therefore what returns can be paid to shareholders, and, of critical importance, what cash is left to run the business with.

Community energy companies need to get to a point where they can employ someone to run the assets they have developed, who can then go on to create more projects. At this stage many do not employ people but remain small and project focused, contracting others to run their operations. However if these companies are going to play a serious role in our energy provision, we need them to be real companies with the staff needed to run them. Invariably they will probably start by using a considerable amount of voluntary time; a sustainable business cannot be reliant solely on that. With Ovesco, our first project provides enough income to pay returns to shareholders, liberate a percentage of capital every year, and employ someone to manage the projects part time for the 25 years of the project. There are not many companies with that level of certainty in their income streams!

So in this section we will look at the various aspects of planning your new social business.

YOUR PROJECT

To be able to get the planning of your first project underway you will need to gather a range of information about it.

- Capital costs – how much will it cost to build the system? Quotes from contractors are the simplest place to start.
- Permissions – if you need any licences to develop the project, how much will they cost? This could include the cost of connecting to the electricity network, or the cost of a planning application.
- Consultants' fees – you may need to engage consultants to get the project to a point where it is ready to build; these costs must be factored into the business plan.
- How much energy will your system generate over the course of a year? Your contractor may well give estimates or predictions, which you will of course need to verify.
- How will the energy generation generate income? Through direct sale of electricity, feed-in tariffs, power purchase agreements, heat metering or other mechanisms?
- What will the ongoing costs of the project be? Meter reading, fuel delivery and management, lease or licence fees, maintenance, equipment replacement, financial management, and insurance all need factoring in.
- Timing – there may be specific factors determining when your project needs to be or can be completed. These could be constraints imposed by the need to access support tariffs, or could be that it is difficult to build the project in winter. These need factoring in.

With this information gathered, you will be able to start working on a financial model for it. The model will be the heart of the business plan, and will also inform you which of the projects you are looking at are the most viable and therefore where to start.

DEVELOPMENT COSTS

There will probably be development costs in developing each project you undertake. Often these are the hardest costs to fund, because if your project does not have all the right permissions people are often reluctant to back you financially. The development stage of the project is, of course, where the risks are highest, and the risks involved will vary according to the technology you are planning to use, as well as the scale and complexity of the project. Raising this initial capital may be something your team can do from their own funds, but if not, it may be necessary to go to a small number of potential investors, offering them better returns on the capital they put at risk during the development process. It might be that local people are willing to invest at a higher risk because they really want to see the project happen. Aspects that may need to be funded during the development stage of a project include the following:

Studies and design work

If you are planning to build a large-scale wind turbine project for your community, you will almost certainly need to carry out a study on the site to assess the wind speed. This may involve putting up a mast to the hub height of the turbine for a period of months so that an accurate prediction of the system's output can be given. Some studies and design work will be needed for many of the technologies and projects. Each technology has different requirements in terms of studies, from structural to environmental, and they obviously come with different price tags.

You may need to undertake design work to draw up exact designs for the elements of your project, be they CAD drawings for solar PV layouts, or foundations for a hydroelectric scheme. Some of these elements may have to be submitted as part of an application for permission.

Legal costs

If you are planning to lease a roof or a parcel of land on which to site your turbine or solar panels, there will be costs to pay to get the lawyers to draw up and negotiate the appropriate licence or lease and option documents. Often when negotiating leases you end up having to cover the legal bill of your potential host also, so it is worth agreeing a cap on fees before you start the process. You may also need to spend some money on the contracts relating to the construction of the plant, and any power purchase agreement you intend to put in place with the energy user or purchaser. The larger and more complex the project, the higher the legal costs are likely to be. Finding a local supportive lawyer who will work on the project either on a *pro bono* basis or on a success only basis might make all the difference to getting off the blocks or not.

Permissions

For all electrical generation technologies you will need permission to connect to the local grid infrastructure (assuming you are not building your own). To do this you will often have to pay a connection deposit in advance to connect to the network. For large projects this can mean that a significant amount of money is placed at risk, until all the other permissions are in place. For smaller projects there may be no fee to connect the project.

In many countries, and for many technologies, planning permission will be required before the project can be started, and once again this can be a time-consuming and costly process for larger projects. Some technologies, like rooftop PV and solar thermal systems, may often be constructed without the need for planning permission. In the UK, the other end of the spectrum is perhaps building a medium or large wind turbine, which can take a year or more to get through the planning process and cost £100,000 of capital that will be at risk.

Community consultation and engagement

Ensuring that the rest of your community support your plan and are willing to buy into it may cost money for publicising and holding events, putting information in the local media, posting letters through your neighbours' doors, and so on. The how and why of community consultation is covered in the section on marketing and promotion.

CAPITAL EXPENDITURE

The bulk of the expenditure for which you will need to raise money is to build the system. Ideally, you will contract one provider that can build all aspects of the project, the mechanical installation including all the costs. On more complex projects it may be that you have multiple contractors to engage in order to make it work. Specifying what it is you want a contractor to quote for is often the first part of the challenge, and working with a contractor to help with specification on early projects may reduce the risk of incorrect specification. Purchase of capital equipment will probably be the largest expenditure that your business has to make, so careful management of the costs here will be required for the project to be a success. Avoiding cost overruns in the construction of the project and unexpected costs cropping up will be part of the project management process that you undertake in getting the project construction ready. You will need to identify a project manager in your team and a number of people to handle the various aspects of the project you are hoping to deliver.

INCOME STREAMS: FITs, ROCs, PPAs, NET METERING ETC.

How the system you construct generates income can take several different forms:

Electricity-generating systems

Total electricity generated
In some areas there are financial incentives to support generation of renewable energy. Most commonly called 'feed-in tariffs', or 'FITs', these incentives pay a fixed price for every kWh of electricity produced by the system, whether it is used on that site or not. To calculate the benefit from this sort of incentive you need an estimate of the total amount of energy generated by the system over the course of a year. If there are any seasonal variations in output, they should be noted because they may affect the cashflow of the project and therefore your

company. Feed-in tariffs, or other similar incentive schemes, normally have a fixed-rate tariff that is sometimes linked to inflation, and that runs for a fixed number of years once the project is constructed and registered.

In the UK we have two schemes like this. One, called the Feed-in Tariff, supports installations of technology of up to 5MW in scale, is index linked and pays out for 20 years once the system is commissioned. The other is called the Renewable Obligation Certificate (ROC) scheme, and provides a premium payment per MWh for larger renewable energy systems. This scheme is a tradeable mechanism and is more complex than the feed-in tariffs, but it still provides a fixed period of returns for the energy produced by the system. Generally these schemes have 'bands' with different tariffs paid for different scales of installations and types of technology, with the aim of providing a stable return for each different scale of installation. Tariff payments are generally paid in arrears, and depend on the size of the system. This may determine the size of your project – to get the optimal return on a site.

Use of electricity
If your system is sited on a building or is close to a host who is using the power, rather than plugging straight into the distribution network, it will be possible to charge them for all the electricity they use that the system generates. This is generally governed by a Power Purchase Agreement (PPA), and I cover the concept in the section on legal aspects (see pages 243-244). The PPA will set the price of the electricity provided to the host and the terms of its review. Part of the advantage to the host of allowing the system to be installed on their site is to obtain cheaper electricity. Generally, energy from an onsite generator is offered to a host at a rate that is a fixed percentage below the standard rate offered by the utility – perhaps 25% cheaper. This is then reviewed at fixed periods, and adjusted up with inflation. This means that the host can guarantee a reduction in their energy prices for the life of the system. If it generates all the energy they are using, this could be a significant saving. Estimating how much of the energy will be used onsite can be done by analysing the projected annual output from the generation system against the annual electricity consumption of the host. Sometimes it is possible to get half-hourly metering data relating to the consumption on a particular meter – this is generally possible for high energy users – and that can be plotted in a graph against the output for the year to determine the proportion of energy that will be used on site. This figure will then also feed into the financial model – and may vary significantly from month to month. Once built, the meters on the system will need reading to determine the actual billing from month to month, although this can be done remotely in many cases.

Export of electricity to the grid or network
Invariably, even if a high degree of the energy from the system is used by the

host site, there will be times when power is being exported to the network. This electricity export will be metered through an import/export meter, and will also generate an income. In some cases a PPA will be needed with the network operator, to govern the purchase of any electricity that is exported into the network. This will often be at a lower price than the retail price paid by the host, but it will also still attract a valuable income. In some areas around the world where there are no longer feed-in tariffs, there is instead a mechanism called 'Net Metering'. This means that if you are generating renewable energy on site and exporting it on to the network you will be paid the same amount per kWhr for the energy you export as for the energy you purchase. In areas where technology is cost competitive with the current grid or diesel generated supply, net metering will potentially make the project viable, by ensuring the system attracts a good price for the electricity it exports. In areas where net metering is in place, any electricity used on site or by the host would be charged under a PPA arrangement, with all exported electricity attracting the same value as any imported electricity.

Heat-based systems

Income will be derived primarily from delivering useful heat to consumers. This could be directly from a system such as a biomass plant or a solar thermal system, into a building or swimming pool, or indirectly via a heat network or distribution system to multiple heat users or houses. In both cases heat delivered to the user will have to be metered to determine exactly how much is used. Heat metering is achieved by measuring the temperature of the heat transfer medium (the liquid that is running in the pipes) as it arrives and leaves the point where it is being used. Taking these 'flow and return' temperatures and adding in the flow rate will allow the meter to determine how much heat is being consumed at any one time. This then clocks up a running total of the kWth being used that can be read periodically so that the user can be billed for their consumption. There are pre-packaged metering units available that can be used to monitor the users' drawing of heat from a district heating scheme.

Each of the units has a heat exchange inside it, so that the fluid that is circulating in the district heating network is kept isolated from the heating circuits within the building that is using the heat. Sometimes they come equipped with a GSM transmitter so that reading the meter can also be performed remotely.

To determine the business case for a heat system, modelling will be required to look at all the heat users on the proposed system, and what their heat consumption has been over the course of the year. This will give us a figure for the amount of heat we will be able to bill for (assuming no one radically alters their heat consumption habits). We will also need to factor in the inefficiencies of the system and the losses in transmission that will occur as the heat is transported to its final destination. Designing a heat system to meet the load will be critical, as unlike an electricity generator attached to the national grid, there is no system to take away excess heat generated, so an oversized plant may be wasteful.

In some areas there are special premium payment schemes, set up by national or local government, to encourage the uptake of renewable heating systems. We have one such scheme here in the UK, the Renewable Heat Incentive, which provides a payment per kWth for energy produced by the various technologies and will make the business case for all renewable heat technologies look significantly better. The UK scheme offers different rates for different technologies and scales of technology. Where a heat incentive tariff like this is not available, there may well be other capital grants that will make the capital cost of the system more affordable and therefore make the business viable over a shorter period.

MANAGEMENT AND RUNNING COSTS

Once your system is built there will be a number of costs to keep it running, according to the type of system and its complexity. Systems that have wind or solar as the primary source of energy have one great advantage, in that they don't need refuelling all the time; nature takes care of that aspect. Other systems, such as biomass or biogas plants, will require considerably more in the way of day-to-day management, partly due to the fact that they require a fuel source to make them work. Here I break down some of the key costs involved with running a project once it has been constructed and commissioned, that will need factoring into you business model.

Meter reading

Whatever the technology, meters will probably be involved, to monitor the generation or output of energy from the plant. Some meters we are very familiar with, as most of us in Europe have an electricity and gas meter in our home. They all require periodic reading so bills can be prepared and bills can be submitted. Traditionally this was done by a physical visit, but today many meters have the ability to transmit their data over the mobile network, making it possible to access the data remotely at any point. If this is possible in your area, it will greatly reduce the number of visits that it is necessary to make to your sites once they are complete. Generally this facility will incur a small ongoing annual charge to access the data, but save potentially many hours of time in travelling between sites. Access to generation data on any plant you construct will be essential, as it allows you to identify problems and solve them quickly.

Fuel delivery and management

For the systems that require external fuel to run, such as biomass systems and biogas plants, the cost of supplies will be a major part of the ongoing running cost. It will also be a significant part of the ongoing work required by the new co-op. The size of the system will determine the scale of annual costs involved. If your business had access to woodland or a source of the primary fuel, it may want to take on the wood processing part of the process as well. This may reduce the expenditure on the resource but greatly increase the amount of work involved, and will almost certainly require specialist skills and machinery, which will add again to the capital requirements of the project.

If the wood supply contract is negotiated with a local contractor, the longer the contract that you can get in place the better, as it will ensure that the fuel supply costs are known for that period of the contract. If this was the first five years of the project, it gives some stability to both the project business and the fuel producer, and if you are lucky will become a contract that rolls on beyond the first five years. Finding a good local partner can be a key part in securing a good supply of fuel, and without a secure supply of fuel of the right quality, your project will not work.

Lease or licence fees

If your project is land based, like a field of solar PV or thermal panels or a wind turbine, an annual land rent or licence fee will probably be payable. It will have been negotiated during the option and lease negotiations, and will normally be payable in relation to the land areas – for example, per acre – or sometimes linked to the turnover of the project in the form of a 'turnover rent'. If the fee is linked to a land rate, the contract will normally include clauses that specify when the rental fee will be reviewed and what metrics will govern any increase. For instance, it might be linked to the retail price index or some other metric of the economic environment.

Operations and maintenance

All energy generation systems require some degree of maintenance, and some require much more than others. Anything with moving parts will need more regular routine maintenance to ensure it keeps working smoothly during its life. As with your existing boiler or car, a routine service should keep it working well, but of course won't stop unexpected problems.

If your system is not running, it won't be generating any income, so having a plan to get it put right promptly if it does go wrong is always a good idea.

When thinking about how best to look after your scheme, it may be that you first ask the contractor who will be responsible for building the scheme whether they offer an operations and maintenance (O&M) contract to look after the system on an ongoing basis. For some projects such as wind turbines, ongoing O&M provision will be a crucial factor in determining which contractor to use. Normally the system should come with a warranty that ensures the system will be in good functioning order for that period, but it may be down to you, as plant owner, to notify the contractor of any problems with the system. An O&M contract may well go further than the warranty offering and mean that you have greater certainty about the system's up time, and therefore greater certainty it will be generating. The aspects of an O&M contract can be broken down as follows:

- Monitoring – if your system is not being monitored, you may not know whether it is actually functioning correctly or not. This could be achieved through remote metering, but also by dedicated software interfaces provided by the equipment manufacturers or service providers.
- Routine maintenance – planned maintenance of the system as per the manufacturer's instructions. This can be pretty minimal in the case of solar PV systems, or fairly involved in the case of biomass boilers or biogas plants.
- Reactive maintenance – fixing things if they go wrong within a prescribed timescale, normally with a predefined call-out cost but with any parts as an extra cost.

Whether an O&M contract is right for the project will depend on a range of factors. The technology being used will be a significant factor. If your system is providing a critical energy supply, such as heating for a number of buildings, having an on-call O&M provider who can attend promptly to problems will be very important for providing continuity of service. For systems that are less critical, such as grid-connected PV systems, it may be that in the first two years an O&M contract will not be needed. However, the larger the system, the larger the potential losses from down time, so the more likely that an O&M contract will be needed.

O&M contracts are normally laid out as annual costs, but often broken down into monthly instalments. In some cases they may well include a performance guarantee which gives the plant owner some security that the plant will perform at a certain ratio of its rated performance. This performance ratio guarantee is sometimes backed by insurance, so if there are problems an insurance provider will come in and pay the difference. This policy would normally be put in place by the O&M contractor as part of the contract. For the business plan, a decision will be needed on whether you require an O&M contract in the first year, or whether the installation warranties are enough. If it is needed from whatever time, it will have to be factored into the annual running costs and cashflows. It will be worth sense-checking the offer you have had for O&M from your construction contractor, to ensure that it offers value for money. Sometimes the contractors who are set up to build a project are not really set up to maintain it as well, so an alternative provider may be able to offer better service.

Warranties

The equipment that makes up your system will come with warranties. These are provided by the manufacturer of the equipment to ensure that it performs as predicted and to outline the conditions for it to be replaced if it does not. As with many other products, it may be possible to purchase extended warranties on some elements of these systems. This adds to the capital costs at the set-up of the project, but gives you certainty on the reliability of those elements of the system, and may be worth considering. It may also reduce planned equipment replacement costs. These factors can be pondered in the business planning stage, to balance the risks with the upfront cost.

Equipment replacement

Invariably some aspects of the systems you have built will need replacing during the lifetime of the project. The cost and frequency will need factoring into the financial model. In solar PV systems we generally factor replacing the inverter halfway through the life of the project. Some projects accrue a replacement fund to repower the technology at the end of its life. Consulting with your contractor should give you the information you need to be able to estimate this.

Financial management and billing

There will be an annual cost associated with managing the finances of your new company, administering any payments to shareholders, raising bills, paying invoices, generating management accounts and submitting annual returns. It might be that you need to factor in the costs of a bookkeeper and accountant to ensure that the bills are paid and that you are clear on the current financial position of the company. Quotes from local or specialist practitioners should be informed by the number of transactions that will occur on an annual basis, which may be minimal at first.

INSURANCE

The systems you build will need to be insured against events outside your control. There are a few possible types of insurance that you might consider:

- Insurance on the system itself to provide replacement costs in the event of a catastrophe or event that causes material damage to the system. This is an essential piece of insurance to have, but it may be that for systems such as roof-mounted solar PV and thermal it is covered by the insurance on the building itself. It is one to discuss with the host site to see whether the system can be covered by their insurance. For smaller systems this may be a possibility, but for larger projects independent cover would almost certainly be required.

- Insurance to guarantee that the returns from the system are still generated if the system fails for a period of time. There are often specialist insurance providers for this type of project. Because they understand the nature of the risks involved, they should be far more competitive than brokers who have never encountered the proposition before. Quotes again need to be sought for your financial model.

- Insurance for your staff and directors. You will almost certainly need some standard business insurance like Public Liability and Employers' liability, Directors' liability and to insure you against every eventuality.

INVESTMENT

There are a number of ways to raise the money to finance the building of your project. Your choice will affect the returns possible to shareholders, and the overall viability of the scheme. I cover the options briefly below.

Equity investment

An equity investment is made by purchasing shares in the company being invested in. This is common to all types of company, however they are structured. Many community energy projects offer investment in their co-op as the means to construct or develop their first projects. Generally the shares will be offered with a minimum and maximum investment level, and there will be a set of rules governing how the shares are sold and what rights they confer. Often they are called 'withdrawable shares', because they can be bought back only by the organisation itself, and at the value at which they were purchased, plus any interest or dividend due. Raising the investment is dealt with elsewhere, but what we need to consider for the business plan is the effect that capital raised in the form of equity will have.

Returns

The returns paid out on different projects and from different co-ops vary between 3% and 8%, depending on the project and any other benefits available, such as tax breaks. Returns will be set by the co-op board in advance of the share offer; they may be determined by the 'model rules'. Often returns are offered as 'social returns'. This means that they may be lower than you might expect on a commercial investment offer, but people will often be happy to accept lower returns because they are doing something that people believe in and they are 'asset-backed' (they are to fund the purchase of capital equipment). The rate of return offered to investors is another factor that needs to be generated by the financial model.

Tax breaks

Tax breaks that increase the rate of return for investors are really helpful in making the investment opportunity more attractive. In the UK we have the Enterprise Investment Scheme (EIS) and the Seed Enterprise Investment Scheme (SEIS), which offer 30% and 50% tax breaks respectively and therefore substantially add to the returns received by investors and therefore the attractiveness of the scheme. As is the case in this area, these schemes are prone to changes and tinkering, so these tax breaks above will shortly be phased out, and replaced with another scheme known as the Social Enterprise Investment Scheme – which achieves a similar thing. In the US a similar scheme exists in the form of the Business Energy Investment Tax Credit (ITC). These returns do not need to be factored into your financial model, but will be an added bonus for investors, meaning that in the case of the SEIS for example, for every £10 invested £5 is returned through their tax return in year one. The investors will still get the returns on the full £10, and in some schemes their full original investment returned as well, increasing their overall rate of return substantially. In most schemes the levels of investment per investor is capped at a certain rate – for instance for Ovesco there was an upper limit of £20,000 per investor. In some cases these upper limits and definitions will determine the shape of the project company and possibly the project as well. Some work will be required to find the sweet spot of any schemes in your area, and how to maximise the benefits for your investors and make your project seem even more attractive.

Equity redemption

If the project has a 20-year planned life, it will be necessary to return the equity invested in the project in equal amounts over the course of the period, or at least to accrue the capital so that it can be returned at the end of the project's life. Returning any capital annually will have to be factored into the business model. If the company has not returned a portion of capital during the life of the project, this accrued 'profit' may be liable to corporation tax. If this capital were left in a bank account, it might well not attract a level of interest that would pay shareholders the returns they were expecting. Returning capital annually gives a portion of investors the option of liberating their capital each year, and may also free them to invest in further projects.

Debt finance

Taking out loan finance to part-fund your project is an option used by many co-operatives and commercial companies. It means that less equity will have to be generated from shareholders, and that each shareholder effectively owns more of the project. Debt finance could come from a number of sources, most commonly banks, but it may be that equity investors are also able to lend money to the project on different terms from their equity investment.

Whether debt finance will be appropriate and possible will depend on a range of factors, some of which are as follows:

- Interest rate possible on the debt finance; commercial interest rates may not be as attractive as lower-rate equity investment.
- Term of the loan (the length of time over which it will be paid back). If this is too short it will mean that repayments are too high to be serviced and therefore the project cashflow will not add up in the early years.
- Whether there are lenders used to lending to co-operatives or community-owned businesses, and on renewable energy projects. If this form of investment is perceived as 'new' or 'risky' by a prospective lender, then it will of course increase the cost of the debt.

Often for commercial debt to work on a project it needs to be at a large enough scale to make it attractive for a bank to carry out the required 'due diligence' to check out the details of your plan and verify that they are happy to make the loan. This could mean that anything less that $1 million is not financeable by standard commercial debt finance, but this will no doubt vary greatly in different areas.

Bonds

One form of debt finance is a bond issue, where capital is loaned in smaller amounts and paid back on a set date, with interest. Bonds don't confer ownership or voting rights, and are sometimes transferable between third parties. They may be useful to raise capital with fixed redemption rights from people or organisations you do not want to share ownership of the project. They can offer a way to raise a form of debt finance from sources other than traditional banks. They may also give the company the ability to raise smaller amounts of capital from a wider number of investors, such as is commonly done using crowdfunding.

Crowdfunding

Crowdfunding may be an option for raising capital for your project. Crowdfunding platforms are generally used to raise either equity, loans or bonds, depending how they are structured. Crowdfunding usually implies that people can invest much smaller sums, and this may in turn mean that more administration is required. Teaming up with a specialist crowdfunding provider may reduce the cost burden, as they have the architecture and systems in place to manage investments of this nature. The leading provider in the UK is Abundance Generation, in France there is Lumo and in the US there is Mosaic.

These leading providers offer the ability for individuals to invest money directly into specific renewable energy projects. It offers the developers of these projects a platform to access individual investors, and promote their projects to them. The developers could be commercial organisations or community groups.

Louise Wilson is one of the founding directors of Abundance and shares why she feels it could be such a powerful tool to engage many people in renewable energy:

> The more the world has become developed, the more we have become de-linked from the essentials of life. Of which energy is one. In doing so we do take it for granted and actually that is not a tenable position for us to be in any more. We need to stop taking it for granted. As soon as you start giving people more information they understand it. If you can make it easy for them to get involved then you solve one of the biggest problems you always have with big issues, which is inertia, people's inertia. Combining something that affects you in your pocket with something you believe in is a way to square the circle if you like. Renewable energy has the beauty of producing immediate results so in terms of being able to bring people closer to their money and what their money does, renewable energy is very good for that. It is solving a problem that we would like to see solved. We didn't talk about ourselves as 'crowdfunding' when we were setting it up. Now there is this thing called crowdfunding and there are lots of people piling into crowdfunding because it's a 'thing'. We started with what we would like to achieve. We've used the phrase 'democratic finance'. Democratic finance is getting everybody involved. It doesn't matter where you live, it doesn't matter how much you make, you can be involved. We want everybody involved because the more people that we turn into stakeholders in this, the more we give them a reason to care. We give them a motivation to understand a bit better and we give them incentives to then fight for it because it is impacting them in more than one way.

People can invest as little as £5 into a project run through the Abundance system, so it makes it open to a wide audience. They offer a debenture in the project – which is a form of debt finance paid back in chunks annually with interest.

> The fact that we have delivered access to renewable energy investment as a piece of debt has been really to maximize its useability by project developers of all sorts. So the big professional developers do not want to give away their equity. They don't have to with this but nonetheless they have got the public in alongside them.
>
> It's democratic because we want all types of projects and all types of people involved to spread the work as quick as we can in this race that isn't getting any longer. We are very keen to work with community groups, one

of the constraints that we see for the co-operative movement is that it's difficult for them to benefit from the economies of scale. Hopefully we can help them.

We are interested in achieving scale quickly because we think that is the most effective way to be a proper counter to the vested interests. You know fossil fuels, nuclear, the guys who are already there, the Big Six. What we are doing, it resonates with people.

I left my banking job, which for the most of it had been a very positive experience. However I had become increasingly uncomfortable because I was just rushed off my feet all the time, zooming around the world. I didn't have time to really reflect on why I found this so increasingly unsatisfying. At the same time I wanted to get involved with the environment and do my bit to try and help. But I didn't really know where it was going to be or how. I narrowed it down to renewable energy, because I worked out that for me, I need to be part of the solution. I'm not a good campaigner. I'm not good at telling people what they should be doing because I feel very flawed myself. I'm very glad other people do it because it's a really essential thing. There are some very much more eloquent people than I to do that. But for me it was 'How can I find a way to make it easy for people to do what they would like to do'. So that was kind of what I was looking for and then I met Carl and Bruce who had got Abundance going.

Just a couple of years in and Abundance has already raised £7.5 million that it has invested into a range of renewable energy projects in the UK. Where could Abundance go? "£500 million by 2020; Bruce is talking about a billion. We have got to see."

DEBT VS EQUITY

Raising money via debt has some downsides, in that it has to be repaid according to a pre-agreed schedule, and generally carries a pre-arranged interest rate. Equity investment, on the other hand, is often not subject to any pre-arranged repayment schedule or interest rates, but generally comes with targets rather than firm commitments.

You will almost certainly have different funding requirements at different stages of your project, and these different stages may well carry very different levels of risk. It might be that for the project development phase, cash is raised via equity through the selling of redeemable shares. In order to make this more attractive, these shares could carry a premium if the project is successful in getting through the development stage. Mixing debt and equity may be very useful in later stages in your company's development. When the limits of local, easily available equity investment are reached, adding debt to the mix will mean the equity could be recycled to enable construction of further projects.

VAT / EVA / SALES TAX

Sales tax, such as VAT or EVA, that is applied to any purchases you make in developing and building your project, will have to be factored into the cashflow of the project. It is possible that these amounts can be reclaimed if you register your company for VAT / EVA. This process will almost certainly have to be done in advance of the purchases, so must be factored in early on. It may be that if you can reclaim the sales tax, you will have a difficulty with cashflow for the period before that is possible. You will probably not want to raise more money than you need, excluding the sales tax if it is reclaimable, or you will be left with cash unspent. You may be able to get a temporary bridging facility to cover the period where the final bill has been paid, but the sales tax has not been reclaimed.

Corporation tax

Corporation tax is generally payable on business profits. Details of rates and structures vary from area to area, but if you anticipate generating and holding profits during the course of a business year, then you will need to factor that into your calculations.

FINANCIAL MODELLING

Taking all the above costs into account, you should be able to build a financial model that will tell you whether the project will be profitable or not. There are many such models available that combine a profit and loss account, a cashflow statement and a balance sheet.

The one we built for our first project at Ovesco had an inputs tab that enabled us to change parameters quickly to establish whether projects were viable if various aspects were changed.

The key inputs for our model were as follows:

- Development costs
- Project management fee (internal)
- System capital costs
- Projected energy yield of the system
- Sale price of the energy generated
- Feed-in tariff or other tariff rates
- Lease costs
- Degradation rate of the system
- Inflation rate
- Debt finance term / interest rates and percentage of the total capital
- Equity finance return rate, redemption rate and percentage of the total capital

- Lifetime of the project
- O&M costs
- Meter reading costs
- Equipment replacement estimates
- Tax rates VAT/EVA/Corporation tax
- Depreciation
- Insurance
- Any community benefit fund

The outputs of this model are a cashflow forecast, a profit and loss account and a balance sheet. The model is crucial in determining whether the business is viable, and, if you are building the spreadsheet from scratch, it will be worth having it examined by someone who understands business planning spreadsheets to double-check all the values and formulas. If you are not an expert yourself, it is time to call in your excel wizard.

With this spreadsheet you will be able to determine whether your business model will add up. This model can be used to compare different projects to see which is the best to proceed with, or which one should be prioritised.

STAFF AND OFFICE COSTS

Many co-operative or community energy projects are started with a lot of volunteer input. While this may be unavoidable in the early days, it is something that will have to evolve and change if your business is to be truly sustainable in the long run.

Factoring in paying someone to run the business as it develops is an essential part of the development process. It is important to realise that it may not be possible to generate a full-time role on your first projects, but over time, as you do more projects and have more assets under management, the workload will increase and having a paid person to run it will be essential. The first projects that Ovesco developed gave us the ability to employ someone part time for one or two days per week. This has meant that we have been able to develop further projects and support others to do the same.

When you have a team or staff they will need somewhere to work. It may be in a home office to begin with, but in time the cost of renting and running an office will have to be factored into the modelling at some stage.

INVESTMENT MANAGEMENT

Part of the ongoing management of the company is managing your investors. This can take a range of forms, from the simple process of communicating and updating them on the state of the project and your plans, to paying them dividends and returning their original investment capital. Most companies require an AGM where shareholders are given an update on the company, its activities and its plans. Investors are a big potential resource for a company, as both active supporters and ambassadors, and as a future source of funds. They need to be looked after.

BUSINESS PLAN

The business plan is the written document that brings all the details of the company's plan together. It can also double up as the investment document that will be circulated to potential investors and debt finance providers. It will tell the story of what is planned, who will do it, when it will be achieved. The key elements to be included are shown below:

- Information about the company – legal structure, registration details etc.
- Information about the project – what is planned, when and why.
- Details of the technology, installation, timescales and process for commissioning and accreditation.
- Details on the wider landscape surrounding the project – political and regulatory.
- Details of the permissions and legal aspects of the projects, such as leases and grid connections.
- Biographies of the Board of Directors and their experience.
- Financial information, expected costs and likely income streams.
- Spreadsheet showing ROI and cash flows for the project over the project life 20-30 years.

If the document is intended as the investment document it will also need to include the following:

- Details of the investment opportunity and possible returns.
- An analysis of the main risks involved.
- Rules governing the sale of shares, voting rights, AGMs, etc.
- A narrative about why people should invest, probably covering both the personal benefits and the wider positive impacts.
- How to pledge an investment.

This plan will have to be checked by the board and any external advisors who have been helping during development. It may need a legal disclaimer on the front of it, outlining the risks and covering the board, because they are raising money from the public. The business plan will be the key document for people looking to support your project so they can understand whether they want to commit their money to it. It needs to be well written and easy to read and, most importantly, it must sell the benefits of investment to the individual as well as the wider community.

CHAPTER 30

LEGAL STRUCTURES, CONTRACTS, PERMISSIONS AND COMPLIANCE

To be able to build any projects or raise any money we will need a legal structure in place, contracts and leases and all manner of legal things to be written, negotiated and signed. For many people this can be a very daunting prospect and certainly the first time I looked through a contract I did wonder at what most of it meant. Whilst dealing with contracts can seem to add cost and complication, they are also an absolute essential. Without robust contracts in place you are potentially exposed to many 'what if' situations. Generally people start out with very good intentions, but when things go wrong it is best to have a framework in place that sets out the position of both parties clearly. Contracts done right will achieve that. Of course it will be essential you get advice tailored to your local conditions, and if you have a friendly lawyer on your board or as an advisor this is where they come into their own. Here is an overview of some of the aspects that will need considering in order to structure a project and make it work legally.

COMPANY OR CO-OPERATIVE LEGAL STRUCTURES

Some community energy projects are run through municipalities or through existing groups or companies, however in many cases a new special purpose company organisation will be need to be established. This is required to provide the legal wrapper for the project so that any assets are clearly held and transactions with shareholders, directors and the wider world are clearly governed.

There are a number of different legal structures that could be used for the business depending on what you are trying to achieve and in what territory you are operating. If you are seeking to raise capital from shareholders for your project you will need to use a structure that enables you to do that. Generally co-operative society structures and public limited companies (PLCs) will give you the ability to offer shares to the general public and are therefore common in community energy projects.

Across Europe co-operatives of various forms have been used for the establishment of local energy companies. There is a strong co-operative tradition in the UK, Denmark and Germany that means that these structures were often familiar to people already. In the US co-operative structures are less commonly used for energy projects as the way the tax incentive schemes work means that investors are less able to easily benefit financially and many of the utilities are either co-operatives already or are in public ownership. In those cases people have found other ways to participate.

It might be you end up establishing more than one company or structure to achieve your purpose. With Ovesco we established a standard limited company first as we didn't intend to raise capital through it. It was also quick and easy and this acts as the trading part of our business. We run it effectively as a 'not for profit' with any profits being used to further our aims of repowering our town. When we had a project ready to build we established a form of co-operative called Industrial and Provident Society (IPS) to act as the vehicle to fundraise into and to own the projects once constructed. IPS as a form no longer exists so now we use societies called Community Benefit Societies (CBS)! Now the limited company acts as the management company for the society to manage its day-to-day operations like employing our staff and paying the bills. Many co-operatives that build more than one project will establish a number of societies – one for each project. This gives a clear legal container for each investment, and also allows different projects that are developed at different speeds and offer different rates of return to be offered out to the community.

There are of course pros and cons to all of the various structures; below I have tried to give you an overview as to how they can be used in relation to a potential energy project.

Co-operatives

Many community-owned energy companies across Europe are co-operatives. It may be right for your project too. Co-operatives have principles that are designed to enhance the democracy of the organisation such as one member one vote, that make them a common choice for these kinds of projects. They embody the principle of democratic ownership and benefit to all – so often a good fit with the broader aim.

Co-operatives are not just used for energy projects; according to the UK co-operative movement, "There are more than one billion members of co-operatives worldwide, which between them employ over 100 million people – thats 20% more than multinational enterprises. Co-operative businesses across the world represent over one billion people and have a turnover of $1.1 trillion; securing the livelihood of three billion people – that's half the world's population." That's a lot of people.

The definition of a co-operative is as follows: "A co-operative is an autonomous association of persons united voluntarily to meet their common economic, social, and cultural needs and aspirations through a jointly-owned and democratically-controlled enterprise."[50] That's our energy company alright.

There is plenty of good information available on the different types of co-operatives and inevitably there will be different variations in different regional locations. One form of co-operative that is commonly used in the UK is the Registered Society. They are allowed to issue shares to the public and therefore they were the obvious choice for us at Ovesco. There are various sets of 'model rules' available, which are pre-registered with the Financial Conduct Authority (FCA). Having society rules that have been used before can reduce the cost and time in setting up. It can be achieved in a few weeks when using one of these pre-registered rules. The model rules we used had already been used for raising money using a tax break in the UK called the Enterprise Initiative Scheme (EIS) (see page 222) that means that our investors got a 30% tax rebate on their investment as well. This helps to make our 'social returns' much more attractive.

In the UK, selling any form of debt product to the public is a regulated activity, subject to the Financial Services and Markets Act 2000. Crucially, co-operative and community benefit societies are usually exempt from these regulations that, coupled with their democratic nature, makes them a perfect vehicle for use when fundraising to develop a project.

If you are establishing a supply business then it may be establishing a 'consumer co-op' works well, where the customers effectively own a share of the company.

Community Interest Companies (CICs)

Community interest companies are organisations set up with the objective of meeting a community interest, so could be used for an energy project. There are of course different forms of this around the world, and in the UK the CIC is a relatively new legal form. CICs may be registered as public limited companies, or as private companies limited by guarantee or by shares (see below) and are subject to company law but include two additional features as follows.

The assets of a CIC may only be used for the benefit of the community and may only be distributed to another body with similar objectives such as another CIC or a charity if the business were to close. This is described as having an asset lock. The organisation must also be able to demonstrate that it will meet a 'community interest test' for it to be registered in the first place.

Because of these special features, CICs are widely recognised as being a model for social enterprise and may be useful if you plan to apply for funding from bodies that seek to fund social enterprise. A CIC is subject to company law but does not offer the protection of the co-operative principles. Although

---

[50]   http://ica.coop/en/whats-co-op/co-operative-identity-values-principles

CICs can issue shares they are very restricted in their use and are not generally used for community energy projects.

Public Limited Companies (PLCs)

Public Limited Companies are able to offer shares to the general public, and are the structure used by many of the large corporations traded on the various stock exchanges. Traditionally a PLC's stock can be acquired by anyone and holders are only limited to potentially lose the amount paid for the shares. PLCs are being used in community energy projects such as those set up by Abundance but interestingly enough they are not using them to issue shares to the general public, but instead a form of debt called a debenture. Generally PLCs are more costly to set up than the other structures, with annual running and compliance costs being higher as well. They do not have any of the principles embedded into them that IPSs or CICs have, so may be harder for community groups to use effectively and ensure that the assets they develop remain in community ownership.

Abundance use this model to enable the widest possible participation in their projects, giving a regular income, but leaving the control of the organisation with the developer – be that community or commercial organisations.

Private Company Limited

The limited company is one of the most popular and widely understood legal forms and it is used by not only standard businesses but also by co-operatives. In fact we set up a limited company first for Ovesco, as it was quicker and easier and we had no need to raise share capital from the local community. It is familiar to the majority of advisers, professionals and funders. In the UK a limited company is registered under the Companies Act 2006 (the Act) that sets out what a company can and cannot do.

A private company limited by guarantee cannot issue shares, instead each member guarantees a certain amount, usually £1, in the event of the company being wound up with outstanding debts.

A private company limited by shares can issue shares, but is prohibited from offering shares to the public. It can in some circumstances issue shares to certain types of investors if it develops an investment document and has it appropriately signed off.

In many cases, a limited company can be set up in as little as 24 hours. However it does not offer the same built-in protections around assets and governance of companies established using the co-operative principles.

There are a range of reasons why it might not be best to go this route. It worked for us but has potential pitfalls when it comes to accessing grant funding, and being seen as a 'community group'.

Limited Liability Partnership (LLP)

The limited liability partnership form (LLP) provides limited liability for its members with the flexible structure and tax advantages associated with a partnership. Membership of an LLP can be made up of individuals and/ or corporate bodies and has to have at least two members. In some cases, a governing document is drafted but it is not legally required. The LLP can be considered as an alternative to a private company. The LLP has become a popular legal form as a vehicle for a number of organisations working together to further a common aim e.g. tendering together for a contract. It may be useful in structuring joint venture projects where more than one company needs to formally participate.

GOVERNANCE

Whatever structure you opt for, how it is managed and directed will be crucial to achieving your aims. If you have taken investment from shareholders it is important that there are processes in place to ensure that all parties' interests are looked after transparently and fairly.

The 'model rules' are what govern some of the legal structures outlined above. They basically set out the governing structure of the organisation and will cover things like the rules about holding annual general meetings (AGMs). They may also cover issues like rotation of board members and election procedures.

Having an AGM for members or investors to be updated on the project and the state of their investment will be important to keep them on side and confident in your company. It will also provide an opportunity to tell them about future projects.

Being able to give clear information as to how decisions were made to your investors is an important part of running the company, so well minuted directors' meetings are essential.

LEASES AND LICENCES

If you are installing a piece of renewable energy technology on land or a building that you don't own you will need a lease or a licence to put it there. The larger and more complex the project, the more complex and expensive the lease or licence process will be.

Leases essentially are a rental agreement that permits your company to use the land or building as a site for your systems. The licence has similar purpose but confers a right to carry out an activity or use an asset. Leases are generally stronger in their legal protection as they are linked to the property not the individual but more costly to put in place.

Any lease or licence will govern the terms of the arrangement between the property owner and the company regarding the siting of the equipment and any access rights to maintain it. It may take quite a while to get this in place, and when building more complex projects it is common to put a 'Heads of Terms' in place first, that outlines the terms of the lease once negotiated. This is a document in plain English that describes the deal – later to be turned into legalese as a lease or licence. The heads of terms may have an exclusivity clause binding both parties into the process for a fixed period of time. This is desirable if there is significant development activity to undertake to make the project happen.

The heads of terms and the lease or licence will almost certainly cover some of the following:

- Term of lease or licence – how long the agreement runs for.
- Terms of remuneration for the land or building owner – how much will be paid for the lease and when it will be paid. This could be a land rent value or it could be a nominal peppercorn rent if the benefit to the host site is in fact reduced price electricity or heat provision.
- Review dates and process for the lease.
- What rights are conferred to the company taking the lease – this might be access rights, or the ability to fix your system to the building.
- What happens if either party becomes insolvent or ceases to trade.
- What happens at the end of the period – it may be you have to remove the system from the landlord's property, or that it is given to them at the end of the lease period.
- Where you are supplying electricity to a host site it may cover the terms of the supply agreement, however this is often kept separate in a 'Power Purchase Agreement' or PPA.
- Definition of responsibilities for the upkeep or maintenance of the building or land that hosts a system.
- Provisions if the host were to sell the property that your system sits on.
- Dispute resolution process.

In my experience lease negotiations will need to be carried out by a legal professional on behalf of the company. Ahead of this it is prudent to put in place a heads of terms so that as many areas as possible are agreed before engaging the lawyers and running up expenses. Often the party seeking a lease will agree to pay the legal fees of the landlord as well as their own to speed and sweeten the process. It is prudent to agree a cap on these fees in advance so that you know what the process is going to cost you before you get started.

On some transactions you may well negotiate the lease straight away, but on others you may first negotiate an 'Option to Lease' before the actual lease. If your project requires significant development work that needs undertaking before exercising the lease and building the project, the 'Option to Lease' will set the terms and conditions that dictate when the lease becomes active. If you did not have this in place you would be potentially liable to pay the lease fee even if your project was not fully developed. To give you an example, to develop a field-based solar array or wind turbine, a planning process needs to be undertaken that can last for six months or even a number of years! It costs a lot of money and takes a lot of work to get this planning submitted and won. Having the option to lease in place means that you are not liable to pay the lease fee until the project is ready to construct. It also gives you the peace of mind to invest the resources in developing the site, knowing that your future tenancy is secured.

Negotiating the lease may well take some time and the larger the project the more costly it is likely to be. If other projects have been completed in your area then there will almost certainly be lawyers who have sample documents that can be re-used for your project that may well reduce the cost of the process. It is well worth seeking out legal advisors who have experience of what you are trying to achieve so you are not reinventing the wheel! This is particularly important for the counterparty – your landlord – as trying to make sure that they use experienced lawyers will almost certainly save lots of additional cost and hassle. There may be intermediaries who have template documents that you can use to reduce the cost of developing the project. If using bank finance in part then banks may mandate use of certain firms.

GRID OFFERS AND CONNECTION AGREEMENTS

Without an offer to connect your system to the local distribution, transmission or heat network, you won't have a business – unless you are planning to build the distribution system as well. Getting the agreements in place with the local grid network operator to connect your system will be crucial. How this is achieved and what form the application process will take varies from location to location and what network you are connecting to. Normally there will be a formal application process and some sort of connection offer once an application has been approved.

In some areas connection for renewable energy sources will be given priority and preferential treatment, however in most they will compete with traditional forms of generation.

In the UK, there is a great deal of competition for grid connections and securing a site legally is worthless without a grid connection offer for the installation. To get that offer you have to submit an application to connect. This will normally be carried out in a specified format depending on the body you are applying to. In the UK this body is the 'District Network Operator' (DNO). The application will need to specify the location where you are hoping to make the connection, what type of generation equipment you are looking to connect, how much power it is expected to export and often some details of the design and circuitry. Once the application has been submitted the DNO have a fixed period in which to respond to it. Currently this is 50-65 days in the UK. They respond with an offer to connect to the network. In the UK they actually have a legal obligation to provide you with a connection offer, but it might not be the cost or capacity you were hoping for. This offer lays out the terms of the connection, which can have a range of different conditions attached. It may be a fixed offer with a period in which to accept the offer, and in most cases it will have a cost attached as well as a timescale for delivery. Generally a cash deposit will be required to secure the connection. This may be a fraction of the total installation costs, or a fixed fee, and it is likely to be a substantial amount for a large project. The deposit may be partially refundable which is worth checking. The offer may be 'interactive' which means that you are not the only person vying for the space on the network, so whoever gets there first may bump the others off the connection. It may also have 'contestable' and 'non-contestable' elements to it. What this means is that some bits of the actual connection works can be carried out by other licensed providers and are therefore 'contestable'. Other parts are non-contestable or non-negotiable and the fees laid out will have to be paid to the DNO to make the connection.

Once a project is built it will often need testing or witnessing by the DNO to complete the connection process and get final approval. Release of some aspects of the finance for your project may be linked to this moment of final connection. This point of final connection may also be counted as the date at which your system becomes eligible for any feed-in tariffs so it is a crucial moment. It is of course also the moment when you start generating electricity and therefore the start of cash flowing into your project.

Connecting to a district heating network is unlikely to be something carried out in isolation, as normally these networks will need to be planned to match supply and demand. In developing a district heating network one of the first pieces of work that will need to be carried out is to get the principle agreement of the potential consumers of the heat from the network. This will need to be done well in advance of its construction and will require a connection agreement to be signed by the heat consumer who is connecting to the network. Often the development process gives a hefty incentive to early adopters of the idea, to encourage high levels of acceptance. This incentive can take the form of a discount on the cost of connecting to the network. Any connection agreement will cover:

• Terms of the connection
• Responsibilities of the company managing the heat network
• Likely costs of heat delivery
• Service agreement

Construction of a new district heating network is a complex and time-consuming process, but one that has been achieved in many places where the conditions are right and there is a clear reason for individuals to get involved.

JVs, TENDERING, PARTNERSHIPS

Tendering

Selecting a contractor to carry out the works required for your project may require a formal tender process to be carried out. Tendering is the process by which you assess offers from rival bidders for the same piece of work you want carried out. It is structured so that the various bids can be objectively compared against each other. It might be that you have worked with a contractor during the development of the project and that you select them to build the project for you without going to tender. I explore this below, but in many cases you will be seeking best value and going out to a number of parties will potentially achieve this. Before you are able to issue a tender and seek bids you will need a tender document. This document will lay out the terms and conditions of the tender, as well as any criteria for the organisations that are submitting responses. It is often common to issue a 'Pre-Qualification Questionnaire' or PQQ to establish some basic criteria about the possible companies who are keen to tender for the work. This may go into details such as:

- Financial standing
- Details of the company
- Numbers of employees
- Experience of similar projects
- References
- Technical capability
- Certification and accreditations
- Policies and track record regarding things like health and safety

Following the receipt of responses to the PQQ, it may be that a smaller number of companies who satisfy the criteria laid out above are then allowed to submit a full tender response. The tender document will need to lay out in detail the specifics of the project, and any criteria that need meeting. These could include:

- Design of the project
- Materials specification
- Definition of the role and scope, e.g. who will be responsible for submitting the various accreditations required
- Access arrangements
- Any time deadlines that need meeting
- Any performance requirements

- Operations and maintenance arrangements
- Warranties and workmanship guarantees
- Price

For many groups new to project development it may be difficult to get some of these details drawn up without access to a specialist technical advisor. There are now many bodies all over the world seeking to assist communities in these endeavours and they may be able to help. Or it may be something that you are able to call in from your network or something you are able to get on a success fee basis. However it may be more practical to work closely with one provider from the start and use them for the construction of the project. There are various ways to approach doing this which mean that you can draw on the advantages of working together whilst maintaining independence in your decision making.

Joint ventures and partnerships

Partnerships between community groups and specialist commercial companies can be a very effective way to get projects developed. For communities who lack the specialist technical or financial knowledge, partnering with a company or specialist in your local community or from further afield that does have the required knowledge can speed the development of the project dramatically. Invariably the development of a project will require an investment of time and resources from your partner so they will want an assurance that they will be the contractor chosen to build the project if development is successful or else be remunerated for the effort that they have put in. For the community group, they will need to be certain they are getting value for money in the process and spending investors' funds in an effective way. There are a number of options to ensure this:

- Reference pricing with competitors – as the project gets close to delivery, check that the prices your partner has quoted are competitive by seeking comparable quotes from others. Inevitably someone will be able to provide a cheaper price at that point but you almost certainly won't get the quality and service from the cheapest offer.

- Transparency – using an open book approach is one way. This means that both parties are aware of the costs involved and the margins being made. It could be possible to have an agreement in place with the contractor that they will make a predefined margin on the project. In the case where the costs change during the project any upside or cost reductions could be shared between the two parties and vice versa. This incentivises the contractor to firstly give a price they can work to, but also encourages them to make cost savings along the way if possible.

Of course all this information will need to be kept confidential to protect both parties in the transaction.

- Setting a target price – if you know what you can pay to make your project add up financially then simply provide your contractor with a target price at an early stage. They can then work to that price in the certainty that if they hit it or better it then the deal will work for all parties.

In my experience delivering good projects is often reliant on the quality of the relationships involved. Yet often people tend to become defensive, secretive and adversarial when choosing a contractor. Any project you build will be with you for many years. The relationship you have with the company that built it will also stay with you, and they may be crucial to keeping the technology running and your returns being paid, so choose your partners wisely and treat them with respect. Beating contractors down on price relentlessly will only serve to get a system built with less care put into it and potentially sour an otherwise fruitful relationship. Choosing contractors simply because they are the cheapest will not get a quality system or provide you with a long-term partner moving forwards. If prices are substantially cheaper from one contractor then there is generally a reason and it is often because some aspect of the job has been missed, or underestimated. Most of us would not consider spending large sums of our own money on the cheapest option if we were installing something on our homes, and the same applies here. A long-term investment such as a renewable energy generator needs to be well built and the cheapest quote won't achieve that.

Often working in partnership between two organisations is a fairly informal process, but it may well be prudent to put a document in place which governs the terms and conditions of the relationship. This could be a simple 'Memorandum of Understanding' (MoU) which lays out the objectives as well as the roles and responsibilities of the two parties and which is then signed off by both parties. It is often really helpful to put something like this in place as it means that both sides have clarity on what they are involved with, what is expected of them and what they can expect in return.

There are quite a few ways in which communities can work together with commercial project developers to realise renewable energy projects where the commercial partner is in the lead:

- Project developers sometimes sell a portion of the completed project that they are developing to a community group to gain their support and acceptance of the scheme. In principle this sounds like a good idea, however it may not be easy to achieve in reality. It may not be simple to 'divide up' a project to give the community a portion of it, and the two parties may not be an easy match. Commercial developers approaching a community will invariably be moving at a different speed to the local community and there may be a challenge in aligning all parties to achieve the desired end result. However where the scheme can be 'split' it can be a great way for commercial developers to assist local communities in developing and owning their own project. The moment people become involved financially things are generally far more acceptable.

- A project developed and constructed by a developer might be sold in its entirety to the community on completion. This can be a very simple way to work for community groups as it removes all of the development and construction risk from the project. These two risk areas are often significant hurdles as to get over them investment is required with no guarantee of success. It may have negative implications for some of the investment tax breaks, but funding a constructed or fully consented project is a much simpler and less risky affair. It also removes the complication of joint development or ownership.

- A community might invite a commercial developer in to develop a project on their behalf, in return for a share in the project company. Establishing a new legal entity part owned by a commercial company and part owned by the community might be the perfect way to generate a situation where both parties are incentivised to make it work. For communities wishing to develop complex projects that would require commercial partners to make significant investments this can be a perfect vehicle, tying both parties into a long-term relationship where they both participate. It may take significant time to find a structure that works for both parties.

For commercial project developers, allowing the community to participate in a scheme may make the difference between it happening or being blocked. It goes well above the standard consulting that occurs when many schemes are being planned, but will inevitably result in a much stronger project for both parties moving forwards. Some of the most interesting projects combine commercial interests with those of the community to develop things that would be impossible without the trust that comes from such collaboration. However the more parties involved the longer and more complex the whole project will become.

CONTRACTS AND AGREEMENTS

All of the works a project needs to go from an idea to a completed installation will need contracts governing them. If you are selling energy or services direct to consumers, that will need contracts also. If you end up employing a staff team, they will need contracts as well. Contracts are there for all parties to ensure that their rights are protected and that you and your fellow directors are covered and working within the law. Here are a few you might need:

Memorandum of Understanding (MoU)

An MoU might be written up between two parties trying to develop a project together. It can be a simple document outlining what you are trying to achieve and the parts the two parties will play in getting there. It will also outline what both parties stand to gain from the collaboration. It might be something you can write up yourself or, if it is a complicated project, something that you need professional help with. It will help both sides in clarifying what they are hoping to achieve and what they are committing to put into the process. I have found it useful to put MoUs in place as it makes sure all parties are clear on what they are trying to achieve and gets them to sign off on it. Whilst it is not a watertight form, it can be useful to set a clear direction.

Non Disclosure Agreement (NDA)

If looking to develop a project with another party, you may have to share sensitive information with them. It is prudent to put an NDA in place to ensure neither party shares anything that the other wants to keep confidential. There are standard forms of words for such agreements that should be readily available on the internet or from any commercial lawyer.

Construction

If you are building a system, the contract will often operate under an 'Engineering Procurement and Construction' (EPC) contract. This contract governs the terms and conditions of delivering the project. These can be lengthy affairs covering the wide range of issues that arise during the construction phase of a project. They generally cover areas such as:

- The parties involved
- What will be done and by which party
- What materials and specifications will be used
- When the project will be completed
- Key milestones and payment terms
- Dispute resolution and termination of the contract
- Procedure regarding variations to the contract
- Any warranties or guarantees (inc. performance guarantees)
- The total price and the cost of any variations to that price
- Retention – often a portion of the contract

Once the process of developing a project and agreeing a price with a contractor is complete, negotiating the contract to get it built may take a surprising amount of time! Often contracts are full of rather strange phrases that seem to make no sense to someone who is not conversant in legal terms; understanding what it all means can add time to the whole process.

ROCs, RECs, FITs, RHI and PPAs

If you haven't worked it out already there seems to be a love of complex acronyms in the world of renewable energy. These ones all relate to the income accrued from the energy generated by the project. If the project is applying for Feed-in Tariffs (FITs), Renewable Energy Certificates (RECs) (US) and Renewable Obligation Certificates (ROCs) or any other incentive schemes that support the generation of renewable electricity, it will almost certainly have to go through some sort of accreditation process to be able to access them. In the UK we also have a scheme called the Renewable Heat Incentive (RHI) that supports the generation of heat from renewable sources. This will also require an accreditation to get registered. Getting projects accredited will normally involve submitting all the details of the scheme to a central register or body administered on behalf of government (in the UK Ofgem). This is often a fairly technical task that will need input from whoever constructed the system for you. It can take some time to get these accreditations but once complete your project will be able to collect payment for the units of electricity or heat it has generated. Often you are not actually paid by the scheme itself, but have to collect the certificates or units and sell them to a third party such as one of the utilities, who are obliged to buy them. Often a certificate will be issued to prove successful registration on the scheme for each generator. In many cases such schemes will have a list of approved equipment and contractors that have gone through some sort of vetting process to ensure their quality or ability to carry out the installations. It is essential you ensure you are using accredited materials and contractors to access the tariff payments.

Often these incentive schemes have deadline dates that need to be adhered to to attract a certain rate of compensation for the project. This varies from scheme to scheme – with some suffering an annual reduction in compensation rates or 'degression' and others suffering a monthly or quarterly one. There may be the opportunity to fix the tariff rate once you have certain key aspects of the project developed.

If your system is selling electricity to the site hosting the project or to the wider network you will have to put a 'Power Purchase Agreement' (PPA) in place. This will govern the terms and prices of electricity sold from your system. It might be you have two in place – one for all of the energy used by your host on site and another for the energy you export to the network – selling to a third party.

Selling reduced rate electricity to your host or landlord may well be one of the attractions for having the system installed in the first place. It is common for the PPA with the landlord to feature a discounted electricity cost when compared to the cost of importing power from the grid. Sometimes this is simply a percentage reduction in costs when compared to power purchased from the existing supplier. Depending on the project and the financials this could be between a 15% and 100% (i.e. giving away for free) saving on the electricity costs they had been paying for that chunk of their supply. The use of total generation, import and export meters will be needed to determine the billing. It might be that the landlord can use all of the generation or only a fraction of it and this will impact the business case. There may well be an annual cost for the use of the meters.

If there is no onsite energy consumption where your generator is located then all of the electricity produced will be exported. In many cases a portion of the electricity will be exported. A PPA with a third party will be put in place to sell the electricity onto the grid. This electricity will then be sold by the third party to consumers. The PPA governing this will be similar to the one signed up with a landlord or host, but it will generally be a shorter term and at a lower rate. PPAs with commercial electricity supply companies will often be at a wholesale price and will generally be for one to five years in duration. At the end of the term a new PPA will need to be negotiated. This may be a good thing if wholesale electricity prices have gone up and a higher price for the sale of electricity can be achieved.

Supply contracts

If a fuel source such as woodchip is required to power the project you are developing, a supply contract will be needed to ensure there is a good source of the right quality fuel available. It may take some time to put this in place and needs work well in advance of the construction of the project to ensure that supply at the right quality and price is available.

TAX CREDITS, SEIS ETC.

If part of the offer to potential investors is a tax break such as the Social Enterprise Investment Scheme (SEIS) in the UK, the project will need to be structured to ensure it complies with any rules and conditions of the particular scheme. Often an advance accreditation will need to be gained to be certain that this can be offered and yet another load of forms will need to be filled in to achieve this. It is worth checking early on in the process of developing any project what the process for applying for advanced assurance is and how long it might take. There is nothing worse than having a developed project but not being able to attract investment because of an uncertainty over this. In the UK it can take about four to eight weeks to achieve this.

PLANNING PERMISSIONS

Before you can build any projects, you may well need planning permission to do so. For some projects this can be a complex and protracted procedure to achieve, and for others it may not be required at all. There is normally a predefined process for your local government office as to how to apply for permission to build a project – the same as there is for building a house. Some schemes won't need any planning permission at all – like most domestic and small-scale commercial solar PV here in the UK, which is classed as a 'permitted development' so you can just get on with it.

In the UK, when planning a large project it is a long and fairly complex process. It often requires a range of studies to be carried out on the site and surroundings to assess the impact visually and environmentally of the construction of the project. Planning can become a very contentious issue and people are often very sensitive about what is going to be built in their neighbourhood, so it is essential this is handled carefully.

For a large project like a solar farm or a wind farm, it is common to have to submit a range of studies on the piece of land you are proposing to use. These might include:

- Visual projections of the completed project – from different locations so local people can see what it will look like in the landscape
- Flood risk analysis
- Ecological impact analysis and environmental enhancement plan
- Analysis of archaeological and historical sites locally
- Construction plan including traffic plan
- Detailed scale drawings of the site
- An analysis of the project's wider impacts and effect on the farming business that is hosting it
- Details of any community offer or fund that will be created by the project
- Noise and aviation studies for wind turbines

It is also a good idea and in some case mandatory to hold some sort of public consultation exercise in the process of working the project up to build support for your project and educate. This might involve a public meeting where people can come and express their views and ask questions about the development, or simply a mailshot to local residents letting them have details of what you are proposing.

Submitting a planning application for larger projects can be a very time-consuming and costly process. It is a process in two phases: carrying out the necessary studies and consultations, and submitting the application to the relevant authority. The process of developing the relevant studies will need experts in the various fields. Once plans for the project are drawn up, experts can be brought in to carry out the various studies required. This will of course come at a cost. At least half the cost and in the case of some technologies up to 95% of the cost of submitting a planning application will probably be to fund the development of the application itself. The other half of the cost will be a fee payable to the local authority to actually submit the application. You can often book a pre-submission meeting with the planning authorities for a small fee, where you can discuss your application before you finally submit it. This can help by ensuring you have covered all the areas the planning officers expect to see covered in the application, so is well worth doing.

SHARE OFFER DOCUMENT

Once you have your project developed and the permissions in place or at least underway, you need to consider how you will attract investment or raise the cash! One common way is through the issuing of shares or debt and to lay out the terms of your offer you normally write a document that can be released to the general public. This document is called the 'Share Offer Document' or sometimes the 'Information Memorandum' (IM). It lays out clearly for all interested parties a range of information about what is proposed. It might be it is a development from your business plan and it is likely to include:

- What is being proposed
- Where it is going to happen
- The costs involved
- Key milestones such as the closing date for raising finance
- The team behind the project
- What is in it for investors
- Any partners or other parties involved
- Details of the investment process and redemption process
- Information on any tax reliefs (like EIS)
- Disclaimers and risk warnings

In some areas it may be required that the IM is signed off by a legal professional before being released to the public. It is advisable to get support from someone who has been through the process before, particularly when it comes to the disclaimers, risk warnings, checking compliance with tax reliefs and making sure it is legally sound. It always seems a bit daunting when your lovely document becomes wrapped up in warnings about how any investment could result in capital being lost, but it is of course essential that people understand that before they invest.

YOU ARE READY!

It took us nearly a year of hard work to finally get all the pieces of the puzzle in place for our first project with Ovesco. We had the lease negotiated, the EPC contract in place, the offer document written, and the marketing out there for our launch, all in all a huge amount of work. Then we found out that the UK government planned to reduce the feed-in tariff that currently made our project viable in just two months' time to less than half of what it was.

The changes took us and the whole solar industry by surprise and meant that we suddenly had a rather scary deadline to work to. We had to raise £350,000 and then build what was at the time a sizeable project in a matter of weeks. Our launch event was two weeks on so effectively when we launched we would have six weeks straight to pull off the fundraise and build the project.

The whole team took a deep breath. It placed a massive level of extra stress on the whole process ramping up the pressure on everyone involved. It is nerve wracking enough launching a new company in a very public way in your local community. Seeking not just their approval, but their cash as well to make the vision a reality. Will anyone actually want to invest? But doing it with a deadline that could jeopardise the whole thing looming large as well made it feel pretty risky. We had some soul searching to do as to whether we had to cancel the whole thing, after so much work had gone into getting it this far. Quite a moment.

The run-up to the launch event was filled with meetings to determine whether we could actually do it, and what deadlines we would have to give ourselves before we committed investment cash to the project. It was a tense period, with lots of meetings and calls to decide our strategy. In the end we decided we had come this far that it would be madness to abort the mission now.

So we set off on our launch with a compelling investment proposition for our local community, and a big deadline looming and no certainty as to whether we would actually be able to meet the targets or not. I am pleased to say that we did.

CHAPTER 31

PROMOTION AND MARKETING

You may love the idea of building renewable energy generation systems for your community, but if it is ever going to happen you are going to have to sell the idea to a whole load of other people as well. Promoting your project is essential to realise the vision. This will inevitably take many forms, from the organic to the highly planned and structured.

It is essential to develop a media and communication strategy around the project to ensure the message is getting out there. For me much of this has to do with telling the stories of change, perhaps around the big challenges we face and the response we have heard about from other communities. Certainly hearing the stories of others in the process of writing this book has given me a great bank of inspiring tales to share with people about what is possible. Using this as the starting point to frame the story of change in your community will help people to feel part of the movement that is developing.

Telling the story of what you are planning is not just about the big picture stuff. It is essential you help those around you to see where they fit in, and how they might benefit. People want to feel part of something they believe in and our projects and companies may well give them that. In the process we are also creating a safe place for people to invest their savings, a decent return on that investment and possibly local employment.

COMMUNICATIONS PLANNING

In my experience the better planned something is the better it will work, and your communications are no different. The more concrete a plan you have, the more likely you are to succeed in selling it to your local community and making it happen. Here are some pointers as to how to start working up a plan:

- Vision: When you are trying to tell a story to the local community it is essential that you first define the vision that you are trying to get across. Hopefully this will be clear from the development of the project – but if not now is the moment to get it clear.

- Audience: Next identify the target audience, both those who will consider the plan as positive, but also those who will view it as negative.

Who will be your potential investors? Where do they go? What do they read? Who will hate the idea?

- Ambassadors: It will really help to build support around the plan if you identify and work with ambassadors who will support and promote your project to their networks. This could be individuals or groups of potential supporters.

- Key Messages: To enable people to understand what you are trying to achieve, it will really help if you develop the key messages that you want to get across. This also means that anyone asked about the project will know what to say. This could be as simple as three short points that form the basis of all the messaging.

- Spokespeople: Who will do the media interviews? Who will stand up and talk at public meetings? It is well worth thinking about who can be your spokespeople. It might be that more than one of the team will be needed, and it may well be worth seeking out some professional support to prep you for looking good in the media. The key messages and practising delivering them will be crucial, particularly if you start doing broadcast media.

- Timeline: Like the rest of your project, the media and communication strategy needs an agreed timeline of activities between beginning and the project going 'live'. Work out where you want to get to by a certain date and then work backwards.

WEBSITE AND SOCIAL MEDIA

The internet is such a useful place for both setting out your stall but also for connecting with a myriad of networks and communities. It is your space to create a platform to showcase what you are planning. A website is a must and it is great that today many of us can create one pretty easily using helpful online tools like Wordpress or other instant web site development platforms. It is the place to give people easy access to the information they need about what you are up to. It will need to include:

- The vision and mission
- Access to all the basic information about your company and project
- Frequently Asked Questions
- Case studies of supporters
- Links to other similar projects that have been successful

Couple a good website with social media and it becomes a powerful tool to spread information and create networks of support around what you are doing. Use social media such as Facebook and Twitter to be proactive and to get people on board.

They can be very powerful for the rapid spreading of ideas and information. They can be used for getting support, raising money, calling people together and building your own community of supporters. Social media turns an otherwise static website into a proactive tool to draw people in. Networks that can be built and deployed at this time are as follows:

- Twitter
- Facebook
- Linkedin
- Youtube
- Google+
- Pinterest
- Instagram

I am sure by the time that you read this there will be some new ones as well!

Content

To have maximum impact on social media and the web in general, creating good content will really help. A picture speaks a thousand words, so they say, and this is true here. Create good images that can be shared, be they photos, or infographics. I love infographics as you can cram so much into them to tell your story – there are even websites that help you make them easily. Making videos is also a great way to tell your story, letting you speak to people directly about your plans. The more engaging your content the more it will be shared, and the more interest you will get.

DESIGN

The company or co-op you are setting up will become a brand, and as such it is going to need a logo and a design that runs through its entire media output. You are going to need to find a designer/supporter within your community to get your brand developed.

Designing and agreeing logos can be a tricky business, and you will need to consider how you get your agreed key messages in the appropriate design materials. This logo or branding will need to feature throughout your media campaign so ensure it is uplifting and motivating.

Then work to strengthen your brand by using the design consistently through social media, posters and publications.

COMMUNITY CONSULTATION

If you are planning a large project it will almost certainly require some degree of public consultation during the development stages. This is a good opportunity to help people understand what you are hoping to do and how it might affect them. This might take the form of a public exhibition in your local town where people can come and see the plans and you can answer questions and get feedback. Large-scale renewable energy projects can attract adverse attention as many myths have been spread about them and the impact they might have on the area they are sited. Meeting people and hearing their concerns whilst explaining what the benefits to them, may help to diffuse any tensions. A public exhibition can be a great way to engage people with the wider issues, and also to tap into the wealth of knowledge that resides in your community. It might be you find supporters with skills you need, or even new potential sites to host future projects.

However you present the ideas to your community, it is essential to create a space where an open dialogue can occur. It is a great opportunity to hear about people's concerns and engage with them, but also a space to do some myth busting.

AMBASSADORS

Ambassadors who support the project and will promote it to their networks can be invaluable in getting the word out. Connecting with other groups and individuals and identifying potential ambassadors in your community and wider afield will be essential to building a community of support for what you are trying to achieve. Looking for groups that share your ideals or are already accessing your target audience has the potential to speed your progress. Provide them with the info they need to understand and promote the project. Use them in advance to sign up and commit funds so you start at launch with people willing to share their story.

Endorsement of your proposal by a local group with a strong supporter base will be a massive advantage when seeking investment or customers. It is worth working with potential allies before launching anything to the public so they feel included and are able to endorse the project. That means making sure they have the information they need to support, endorse and promote your project. Outreaching to influential local individuals who may be interested in investing is also worth doing before you launch. If they are supportive this should give confidence when taking the offer more widely. When we came to the launch of our share issue for Ovesco we already had nearly half the money we needed for the first project pledged beforehand. That made it a whole lot easier to get the rest.

CELEBRATION EVENTS

There is nothing better than bringing people together to celebrate successes and launch new projects.

Launch

Inviting your community in to hear your plans is a real opportunity to get them engaged and enthusiastic about what you are proposing. It is another great moment where you can tell the story of change: from the big picture down to what you are planning in your town, and how all of those present can take part. The launch night for Ovesco was a big night for all of us involved. We had a packed town hall with about 300 people turning up to hear our plans. There was a real buzz in the air as the hall filled and we opened the meeting. We had recruited supporters to make the evening run smoothly, a local restaurant had put on some nibbles, the brewery – Harveys – and our host site had brewed us a special 'Sun Beer' for the occasion. We even got a couple of companies to sponsor the event to cover the costs of putting it on. Having founded the company and been the person with the original idea, I had the pleasure of standing up to tell the story of what we were trying to achieve alongside fellow director, Liz Mandeville. It was a big moment for me, and minutes before I went onto the stage to start the proceedings I bent down to get something out of my bag and had a minor accident. So what my fellow residents of Lewes didn't know that evening was that there was a huge rip in the seat of my trousers as I stood there on stage.

The show must go on, and I just made sure I kept facing the audience all evening, but it certainly added an extra level of pressure! The event was a success, and we had arranged for an energy 'question time' following our talks about the project where we had invited a range of politicians and speakers to answer people's questions about our energy future. At one point we had a grand unveiling of a 'totaliser' that showed we had already got half the money we needed pledged.

Within a few weeks we had raised all the money that we needed for the project. I am certain that the launch had a very positive effect on this, with people leaving the evening feeling like they were part of something.

Organising a motivating and exciting launch event needs careful planning and will probably need a host of volunteers to make it run smoothly. Things to consider when putting one together:

- Venue – somewhere neutral that you will be able to fill on the night
- Promotion – making sure people know there is an event happening that they need to be at
- Panel of speakers – your company directors will need to outline the plan of what you are up to, but you may also want to pull in other speakers

who will increase the profile of the event
- Facilitation – ideally you will have someone that people know to chair the evening and keep it to time and in order
- Q&A session – it's important that people have a chance to ask questions and have them answered
- Refreshments – it always helps if food and drink is on offer
- Printed materials or offer document / pledge forms for people to sign up to support the project should be available on the night
- The structure of the evening – is there space for people to talk to each other and share their feelings about the project as well as sit and listen?

Community celebration

When you are successful it is really good to celebrate, and build on what you have done. Bringing together all of your supporters to network, talk to each other and toast your achievements to date will really build the feelings of support for what you are doing. Celebrating the milestones and achievements is a great way to honour all involved for their part, be they investors, volunteers or you as protagonist.

MEDIA RELATIONS

Getting coverage of what you are up to can be a massive boost in getting your ideas out there or promoting a launch event or share issue. It is a far more effective way to spread the stories of what you are planning than any paid advertising or mail drop could ever be.

Identify key media in your area

There are a whole range of different channels to put out your stories that will get you in front of different audiences:

- Local print media – papers and magazines
- National print media and magazines
- Local radio and television news and current affairs programmes
- National radio and television news and current affairs programmes
- Online magazines and news channels

But it will help to identify exactly who it is you want to target and get the coverage to focus your campaign towards them. For many projects getting the wider local community up to speed on the project will be the main aim and this can easily be covered by the local press.

Local press

Many towns or localities have small newspapers, radio stations and even regional news channels covering the stories and happenings of the area. These are often quite small operations that are only too happy to cover stories when presented in the right way.

Most important is to call to check whether your story is likely to get featured before you submit it. It might be that your chosen media outlet will be able to give you some guidance as to whether they have a packed schedule for the next programme or edition or whether they can feature you right away.

Working out production deadlines for the various media you are hoping to hook will help you to work with them to get the coverage you are seeking. Make sure you are available on the day the PR is released to do early radio shows or lunch time TV broadcasts and that you have clearly worked out the messages you want to get across. If you end up doing an interview for a local TV news programme you may find yourself with the camera crew and reporter for an hour or more. The temptation is to try and give a long-winded explanation of what you are up to, when in reality whatever they film will probably be cut down to 20 seconds, so you need to keep it succinct and to the point.

Draft press release

A 'press release' is a document that details the story of what you are trying to achieve in a simple way. Working out what is new, unique, interesting, or topical about the plan will be really helpful to hook interest. Generally a press release (PR) will cover the following themes:

- What is going to be done and where
- Who is doing it
- When it is planned to happen
- Why it is happening
- Quotes from some of the parties involved
- How people can get involved
- Contact details for who can be interviewed for the piece

Photos tell a thousand stories so if you can take some high quality photos to submit with your press release that will really help any printed coverage to have impact.

Persistence

Once you have sent out your release to your target media, follow up with a phone call to find the right person to speak to.

- Persist – keep trying until you find the right person and get to talk to them about your story

- Invite them for a coffee and a 1-2-1 interview – or interview with key people
- Keep them informed on progress

National and international press

If you have a unique story to tell or something that is very topical or controversial it may be possible to get national coverage. This is great for building your organisational profile and accessing a much wider pool of possible investors and supporters.

Finding correspondents or journalists in the relevant papers or channels who have written on similar subjects and approaching them directly will give you more chance of success in getting coverage than sending an email to the news room.

Often if you get coverage in one national paper or broadcast outlet the others may also feel compelled to cover the story too. So working hard to find those first national opportunities may well open doors to make your story go really big.

Getting national coverage for a local story will be tough, and looking for the angle that makes it relevant and of interest to people across the country will give you a better chance.

With Ovesco our launch took place with the backdrop of sweeping changes to the Feed-in Tariff that made the installation add up financially. We were also the first community energy company in the UK to try and build a PV system using only a local share issue, so we had a couple of hooks that helped to make the story prominent. We also had the help of a seasoned PR firm to manage the messaging and work on securing the coverage, which was great.

Repower Balcombe had a unique PR opportunity in being the first fracking village in the UK, with a big media spotlight on it already. When that team launched the company to the public they managed to get featured on prime time BBC TV telling the story of a community taking control of their energy future and spreading that idea to millions.

If you are planning a longer running campaign or if you are seeking national coverage it is well worth seeking out the help of a professional PR advisor to help. There is a lot of work involved with getting repeat coverage in the press and a good PR specialist can make this process smooth.

Timing and message

With any media, be it broadcast or print, working out a key soundbite or paragraph that covers your top four messages is a great discipline. The temptation when asked a question by a reporter is to ramble on and fill the space. The discipline of honing your message to a short soundbite means you are far more likely to get your message out in the 30 seconds you have. It can be very powerful!

Picking when to make a media noise is something to consider carefully. When you are in full fundraising mode it is great to have media coverage to help the process. Picking one outlet to work with exclusively can sometimes help lend weight to your story.

WORD OF MOUTH

It is the best! I much prefer it if someone I know tells me about something they rate or are excited about. I am far more likely to investigate. There is nothing better than getting people talking about stuff that excites them. If you can generate a buzz around your project that gets people talking to their friends and families then your support will naturally grow. Give people something great to talk about!

INVESTOR RELATIONS

Once you have built a community of investors and supporters you need to keep them up to date with the journey. Keeping people informed as to how your project is developing and changing is essential when it comes to keeping them on board and enthusiastic. You may well need their enthusiasm and support with your next project. This could take the form of regular newsletters during the process, updating them with the latest achievement, be that projects built, money raised or energy generated.

CHAPTER 32

PROJECT MANAGEMENT AND BUILD

To get any project completed, whether it is a generation project build or a community engagement process, it will need management. Many people have the skills needed to put all of the elements in place to take a complex project from an idea to completion, but to begin with we are going to need a plan!

Ideally the plan needs to set timescales to achieve certain tasks for the whole of the project from start to completion. It also needs to state who is responsible for each of the elements, so that multiple parties can feed into the process.

Developing a community-owned energy company or project in your area requires a huge amount of upfront work. Getting that work done in the right order will make things flow much better, and mean that you are more likely to achieve your ambition. It is also surprising how long it can take to get a project developed and starting with a clear plan where possible can help manage expectations as well as giving all the participants clarity on what is expected of them.

One person will need to be responsible for ensuring the project runs to plan – the project manager. It may be that the skills needed are already in the team, great if that's the case. If not bringing in a specialist project manager may be required.

MANAGEMENT OF SUBCONTRACTORS

If the project involves using the services of specialist subcontractors they are going to need management. This means someone has to understand exactly what they have priced to do, and with what materials. Making sure they are building the project in the way that it was originally designed, with the correct materials, on time, to budget, and safely is very important. Once again if you have no experience of managing projects using subcontractors it may be the moment to call on your network for help. The more complex and costly the project, the more detail will be needed.

DESIGN VERIFICATION

Verifying the designs and specification of the project you are developing can be achieved by bringing in specialist advisors. Relying on the subcontractor to produce designs and specifications may not end up with the optimal system being constructed. Bringing in a third party, whilst adding expense, can give comfort to the team pulling the project together, but also the external investors or banks. The larger the project, the more likely it will be needed.

CASHFLOW MANAGEMENT

Financial control of a project during the construction or development phase is also an area that requires some pre-planning and consideration. Cashflow will need to be managed to ensure money is released against milestone achievements, that are able to be verified. Ensuring that the contractor has completed aspects of the work before releasing payment will be crucial. It is common to hold a retention on completed works, sometimes for up to a year, to ensure any defects that become apparent in the first year are fixed promptly. Retentions are often 5% of the total contract sum.

THE FIVE Ps

The more thought and planning that goes in upfront the better the end result. As my father used to say to me, "It's the five Ps: prior planning precludes poor performance". But of course there is a balance to be struck. It is easy to spend all of our time crafting the perfect plan, and yet never getting anything done!

CHAPTER 33

BUILDING A SUSTAINABLE FUTURE

Community energy initiatives often start with hundreds of hours of voluntary labour with a dose of hope and determination. Certainly Ovesco was that way in the beginning and still at this point relies on voluntary directors (such as myself) who take the pleasure of achieving the vision as our reward. Many of them stay that way, and that works for those involved. However this is not something that can sustain and grow the organisations we are building into the future. For them to be truly sustainable, they need to be financially sustainable for those involved. Volunteers need to become paid employees as the companies develop their income streams and assets. Our success hinges on our ability to make these businesses, in service of our communities, also serve those whom work in and on them.

This may just be a case of scaling the activities being carried out until there is enough profit being generated to afford needed salaries as well as the expenses of running the company. Before starting it is well worth thinking through what this might look like – if that is where it is destined to go. Some groups opt to keep their initiatives small and have others manage them. However if you do want to grow a business, what milestone do you need to achieve to make enough money to employ your first employee? That could be a number of renewable energy projects built, or people signed up as electricity customers. You may struggle to get to these milestones in one step, but if you have a plan at least you know where you are on the path. Sometimes applying for grants may be a step along the way to get started, and smooth out the early days when there is no money and everything must be done for free. Sometimes your community will be keen to support you through this early phase, by funding your work. For us with Ovesco we know that when we have about 5MW of projects constructed, we will be able to fund an employee full time and run an office for the next 20 years on the income generated. From that base we will be able to grow and expand further, so our immediate mission is to get there.

Developing a plan to deliver the vision of your group and your community is essential. It is tempting to feel that we need to help our community to go 100% fossil free in one year. However this is never going to be possible. Far better to work on a vision that lays out clear incremental steps along the way to make it possible. Perhaps this is to plan that in 10 years we want to get 50% of our electricity from locally owned renewable energy. We can then look at the pathway to achieving this, and the projects that will need to be delivered along the way. We will also be able to plan at what point our business starts to be sustainable on this journey.

Revisiting and building a strong collective vision with the key people involved and the wider community is an essential part of our success.

Seven years into the Ovesco journey, we finally realised that we really needed to take some time out to reflect on our journey and revitalise our vision moving forwards. We were honoured to win a prize called the Ashden Awards, and not only did this give us a much needed moral boost and some financial support, it also got us thinking more strategically about where we were trying to get to with our project. We used some of the prize money to indulge in some much needed collective reflection.

We spent a series of meetings together with an independent facilitator reviewing what we had achieved and how we felt about it. It is amazing in a small dynamic group how many tensions and unspoken feelings can develop over the years of doing something together, especially that challenges the status quo. For me those meetings were a very nourishing moment for our team to share ourselves in a deeper way: our hopes, fears, and frustrations. A moment to reflect on how much we had brought to each other over the years, reflect on the tensions, laugh and shed a few tears. When we set off on our journey our shared vision was very clear, and over the years it gets clouded. However with just a few afternoons' work, we walked out remembering what a good team we are, feeling valued, and appreciating what each person brings to make it work. Most importantly we walked out ready to work together strongly to achieve our vision. Invaluable.

So take some time to reflect.

Now get on with it!

In a gentle way, you can shake the world.

Mahatma Gandhi

PART IV

Step Forward in Hope

CHAPTER 34

HOPE

No matter what people tell you, words and ideas can change the world.

Robin Williams

For me this really hit home when I was talking at a solar conference in Canada about the solar market in Europe. I happened to see another session at the conference on community energy and decided to go along. We had completed our first project with Ovesco just months before and I was excited to hear what was happening in this part of the world. I was amazed when the first speaker started talking about their project, and then proceeded to hold up a copy of the Ovesco share offer document citing it as one of their inspirations in starting their project.

I found it hard to believe that our little story and our one small rooftop project could have travelled across the world and seeded itself in another community. Not only had it seeded, the idea had been nurtured and taken root. Here were people applying the idea in their town and sharing it with others. Not only was this really exciting, it was also really moving for me, to know that our work had an impact way beyond what we had imagined, or known about. That here was an idea that resonated with people so much it had travelled to the other side of the world and been applied there also.

We live in a time where ideas and information move incredibly fast. Good things can spread so easily. Ideas that work spread themselves, and start to grow naturally as the above example illustrates perfectly. Look at the work of Paul Fenn in creating Community Choice Aggregation in the US (see pages 78-83) – seven years of hard work before a first success, then from nothing to touching one in 20 electricity consumers in the US in 12 years, amazing.

Let us draw hope from the fact that we can so easily share in each other's learning to build the fleet of solutions that we need. New companies held in common that solve energy problems for people around the world, whilst also reducing our emissions, and bringing us together.

No snowflake in an avalanche ever feels responsible.

Stanislaw Jerzy Lec

More and more of us are waking up to our part in this and not only despairing but taking action. We are realising that working in isolation will never solve the issues we face, so we are coming together as communities to find solutions. The growth of communities taking action is also increasing dramatically as these ideas spread. These first mover communities are putting in the hard work, but also reaping the benefits, and these benefits are not only for the planet, but for the long-term health, wealth and wellbeing of their people. From Bangladesh to Brixton those who adopt these new models are locking in positive benefits and strengthening their community in the face of these huge challenges. The communities who embraced wind energy in Denmark went on to birth the global wind industry and reap some of the rewards of its success.

We often talk of our climate problems in terms of reaching a tipping point, or a point of no return. However I feel this concept of tipping points is perhaps even more helpful when we consider the systemic change for good that is occurring in many places. As the penetration of renewable electricity being delivered into the German grid has rapidly inched up to over 28% of their total supply, we are reaching a tipping point moment. This point where the incumbents are openly talking about a radical shift in the viability of their businesses because of renewables surely signifies the moment we have been working towards is being reached.

In fact Germany's biggest utility firm, E.ON, has announced that it plans to split in two and spin off most of its power generation from fossil fuel and nuclear, saying it wanted to focus on renewable activities.[1] Surely a sign that the tipping point has been reached – it's now a matter of time before the others all follow. As people across sub-Saharan Africa realise the power and elegance of the solar light and how it can transform their lives, the deployment grows exponentially and demise of the kerosene light draws ever closer. At some point soon the solar light becomes the norm and the change is inevitable. We are speeding to that point.

The concept of 'unburnable assets' is starting to take root as people from all sectors wake up to the fact that the fossil fuels need to stay in the ground if we are to have a liveable planet. Even the Bank of England has started an investigation into the risk of an economic crash if fossil fuel assets have to be left in the ground.[2] Right now it is not about the scale of the divestments, but more about the loss of social licence experienced by the fossil fuel industry. Inevitably there will be large parts of the fossil fuel incumbency that continue to deny the inevitable and plough on with ultimately suicidal investments and activities, like Royal Dutch Shell plugging on with Arctic exploration. But no more will you do these things in our name.

Reaching a climate deal seems more possible than ever after the US and

---

[1]   'E.ON to Quit Gas and Coal and Focus on Renewable Energy', www.theguardian.com/environment/2014/dec/01/eon-splits-energy-renewables

[2]   'Bank of England Investigating Risk of "Carbon Bubble"', www.theguardian.com/environment/2014/dec/01/bank-of-england-investigating-risk-of-carbon-bubble

China announced a bilateral deal in November 2014. This landmark agreement was announced jointly by President Obama and President Xi Jinping. Without the world's two biggest emitters of CO_2 signing themselves up, the rest of the world would have struggled to come up with a meaningful agreement. It sets new targets for carbon emission reductions by the United States as well as a commitment by China to cap its emissions by 2030.[3] It's the first time that China has committed to stopping the growth of its emissions, and it seems likely that their emissions may well drop before this date anyway.

Efforts to reduce emissions are actually starting to pay off. As the International Energy Agency reported, in 2014 global emissions of carbon dioxide from the energy sector did not rise for the first time in 40 years. "This gives me even more hope that humankind will be able to work together to combat climate change," said IEA Chief Economist Fatih Birol.[4]

For me a huge source of hope is the fact that renewable energy is cheaper than fossil fuels in many areas. The price drops experienced in both wind and solar PV technology as manufacturing and deployment has scaled up is a huge opportunity for people all over the world. In many places solar PV is fast becoming the cheapest form of electricity. As this reality sinks in, why would anyone build a new fossil fuel generator when they will be undercut by renewable energy? What is more, the modular nature of solar and other renewable energy technologies mean that it can be applied at a huge range of scales from 10W to 1GW so that so many more people can be involved in the generation of their energy.

I feel we have reached a time of cross over, where the change we are seeking is becoming inevitable. That does not mean it will be easy to effect, or that it will happen of its own accord. In reality it won't be easy. It's going to require immense hard work from people all over the world. However, the conditions for the change to occur are more favourable than they have ever been. As Agamemnon from Repowering London puts it so well: "We have all the technology and the money we need to do this." We are racing to a positive change tipping point and it won't need that many more of us to join to tip the balance for everyone. This energy revolution may well surprise us all and happen far faster than anyone could have imagined. Many of the communities I have spoken to have achieved their aim of being 100% renewable in less than 20 years which goes to show a huge transformation is possible in a short time.

These ideas are infectious so let's spread them.

People want to have meaning and improve the world. A report by social change consultancy Global Tolerance has found that Britons are not happy

---

[3] 'US and China Reach Climate Accord After Months of Talks', www.nytimes.com/2014/11/12/world/asia/china-us-xi-obama-apec.html?_r=0

[4] 'Global Energy-related Emissions of Carbon Dioxide Stalled in 2014', www.iea.org/newsroomandevents/news/2015/march/global-energy-related-emissions-of-carbon-dioxide-stalled-in-2014.html

with traditional business practices and want employers to do more to improve society and the environment.[5] I think that people all over the world feel that way. The environmental protestors in China who act with much personal risk, as well as many others in other nations' certainly reinforce this view. Let's face it, when it comes down to it we all want to see our children, friends and families succeed. We all want to live in a place that is not polluted, where there is enough to go round, and where there is space for nature.

Locally owned renewable energy projects are a huge opportunity to create meaning and equality and find that purpose once more. Whilst at the same time making things better and doing our bit towards solving the climate challenge in our towns and for our communities. Bringing back the balance, re-embracing the elegance of this amazing planet that we inhabit by copying nature's patterns to power our existence. We have all we need to make this change, now we just need you and me to step up and get organised.

It is no longer enough to march in the streets and click the online petitions about things we are unhappy about. We can't wait for governments to solve this for us. It is time to organise. It is time to build, it is time to create.

To build a new fleet of companies for hope, with democracy at their heart.

To build the new local savings mechanisms to fund this journey.

To create a new 'commons' in the delivery of our energy.

We have the blueprints we need. Now is the time to apply ourselves and speed this revolution. As others have already done in so many places, now is the moment to apply this in our towns and our communities. They don't call it power for nothing – time to reclaim it.

It is time to make the stories that we will tell to our grandchildren. The stories of hope. The stories about how the problems seemed so insolvable, but that we organised ourselves and started. That we took the first step in faith, with friends and neighbours. That people all around the world organised themselves and changed history. It is happening today.

It is time to wake up to this opportunity.

Now is our chance.

---

5   'It's Not Just Millennials: Nearly Half of Brits Want Jobs That Change the World', www.theguardian.com/sustainable-business/2015/jan/15/employers-social-environmental-business-practices-survey

CHAPTER 35

MANIFESTO FOR THE ENERGY REVOLUTION

FOR US

Get our own homes in order
If you can, sort out your own home, business, school etc. Reduce the energy needed to run it and then get generating renewable heat / cool and power.

Organise in our community
There has never been more need for communities to come together and organise themselves around energy. Every community should come together to establish an energy plan to develop and own renewable energy generation as well as reducing consumption in our homes and businesses. There are a myriad of examples in this book. It is not enough just to do this on our own in our own homes, it's time to go out there and organise with others in your community.

Develop new companies for change
You can become an energy entrepreneur. Look at what business models work in your community and then set about making them happen. Build local renewable energy projects in democratic structures that enable broad participation. Build companies that share the benefits and become beacons of hope for all involved. Build companies that stop money leaching from your community into multinational energy companies, but that keep it locally, providing employment and energy, and increasing wellbeing. Look to how you can use any surplus generated to reach and assist those most in need in your or other communities, tackling fuel poverty and helping energy access.

Move our money
Move our money out of the fossil fuel industry into local renewable energy projects and energy companies. This could become the new form of a local savings bank. Where better to put your money than in local productive green assets that you can actually go and see? Much safer than leaving it to the whims of the international stock markets and the global banking sector.

Network and replicate
Once you have achieved a success however small, tell the story. Open source the models and best practice, train others and share the learning. Get your story out there, along with the stories of the pioneers. It is not time to hoard a good thing. Our success in this endeavour relies on other's success. We cannot do this alone, so encourage, mentor, advise, share and help speed the change. We need a new energy company in every town.

Tell the stories of hope
It is time for a new narrative of hope and solutions – we all need to feel there is a chance to solve the problems. Tell the stories of change from communities around the world like the ones contained in this book. Tell the stories of change in your community, and tell the stories of where we will be in five or 10 years, with continued success. Inspire people to step forward out of their comfort zones and get involved. It's happening, and we need to let others know!

FOR OUR LEADERS

Energy saving as a priority
Governments around the world should institute radical energy management programmes to reduce the energy and power consumption of their countries. This could enable us to turn off the most polluting / risky energy generation sources like coal and nuclear plants overnight. What was achieved in Japan with no investment, but massive awareness and top-down leadership, could be copied anywhere, and should be. Imagine the impact if President Obama was on TV tomorrow talking about a radical plan that we were all encouraged to be part of.

Fuel poverty
It is time to address fuel poverty globally, both for those in industrialised nations, who are dying from cold and for those in other countries with no access to modern energy, dying from exposure to smoke. Governments across the world should work together to bring dignity to their populations by enabling these fundamental rights.

Instigate a government-backed zero interest loan package for consumers and businesses to invest energy efficiency and renewable energy measures. This could lock in the saving and make our progress to 100% renewable so much more achievable.

End fossil fuel subsidy
Stop throwing fuel on the fire with fossil fuel subsidies. Level the playing field by removing public financial support for the fossil fuel industries. Enough already.

No new coal

If we are serious about leaving a habitable planet for our grandchildren, then we cannot afford to build any more coal-fired power stations. Coal-fired power stations constructed now will likely become stranded assets anyway as the energy revolution gathers pace. We should be asking for a moratorium on construction of new coal plants globally. Governments should seek to phase out the most polluting power stations first, by the introduction of increasing carbon standards for the generation of energy.

A plan for nuclear

Enough money has been wasted in chasing the nuclear dream. In its current form it is over except for the toxic legacy. The priority now must be on what to do with the vast issue of waste that is spread globally. An international task force needs to be established to come up with a strategy as to how to keep the many coastal nuclear power stations safe with rising sea levels and deal with the waste issue. Support needs to be given to nations that have nuclear accident sites.

A plan for the demise of big energy

We need to start a conversation as a society about how we will manage the demise of the fossil fuel companies and the utility monopolies. This will be inevitable with this transition and these companies currently play a dominant role in our economy and society. Will these companies be able to transform their business models? If they cannot adapt in time does that pose a threat to the stability of our society? We need a plan to manage this change.

100% renewable

Adopt the 100% renewable goal and put plans in place to get there. Whatever organisation you run, be it a school, a small company or a nation, look to set a 100% renewable goal and set steps on the path to getting there. It is achievable, and will become the norm, but people need education to understand that.

Stable policy to achieve clear targets

Put in place stable policy and regulatory regimes to support deployment of renewables to deliver ambitious but practical renewable energy targets. This should be built with cross party political consensus where possible. Give the sector a clear pathway to develop and grow, whilst reducing financial support in a measured way over time. In this way renewable energy can be free of public subsidies in less than a decade. Build a framework around renewable energy and community energy to encourage training, education and innovation in the sector.

Priority access for renewable energy

Renewable energy should have priority access to the electrical grid – in advance of fossil fuels and other forms of energy. When demand shedding needs to occur because of over production, fossil fuels should be turned off first. Community energy companies should be given priority access to the grid above other investors, with a reduced or fixed connection charge, when compared to commercial developers.

Interconnect nations

Neighbouring countries need to be planning and investing now to build electrical interconnections so that power can be transmitted between them in times of excess or need. This work needs beginning now to enable us to reach our renewable energy goals in the coming years.

Tax breaks for investors

Encourage individual, community and corporate investment in renewable energy and energy efficiency by making it tax efficient to do so. Tax breaks are given in many areas to encourage investment. Community-owned renewable energy needs to be assisted with tax breaks that engage individuals and their money. The right tax regime could turn community renewable energy into a great savings product.

Communities able to sell energy

Allow local municipalities and communities to aggregate power purchase in their area and supply to the residents of that area. Becoming a local energy supplier, with democratic or co-operative governance, will transform the dynamics of the power sector, putting control back into local communities' hands.

Incubators for community energy companies

A small amount of support goes a long way for fledgling community energy enterprises. Grants or a fund for communities who are just starting out to get the first steps made will be an essential way to speed the development of the sector.

Storage and smart grid

Funding for research into storage and smart grids is needed. We need to help accelerate the development of the smart grid systems we will need to balance the variable output from renewable energy, and enable more intelligent use of energy by the end users.

North / south funding

Those of us in the industrialised and richer nations should build into our community energy business plans a profit share for those in the developing nations. Our local energy revolution success stories could be the fuel for similar success in communities where access to capital is next to impossible, and access to energy is non-existent. Let's build a fleet of companies that don't only help our community, but share the wealth and learning with those who need it most. Perhaps this could take the form of twin energy towns – where communities from different regions work together in the transition. Let's build wellbeing for all.

Global climate deal

We desperately need our leaders to come up with a meaningful climate deal that sets the world on the path to zero carbon that will provide the impetus and backdrop for communities to make the changes needed to get us there.

Conversion Formulae

m – ft multiply by 3.280

ft – m multiply by 0.308

ha – ac multiply by 2.471

ac – ha multiply by 0.404

m^2 – ft^2 multiply by 10.763

ft^2 – m^2 multiply by 0.093

km – mi multiply by 0.621

mi – km multiply by 1.609

Glossary

CHP (Combined Heat and Power) – Production of both electricity and useful heat in one generation system.

Co-generation – Like Combined Heat and Power (CHP), this means generating usable heat and electricity from a generation device rather than just one of them.

Feed-in Tariff – A scheme to incentivise generation of energy from renewable sources, by giving premium payments to the owner for the output from the system. First established in Germany now in many countries worldwide.

Lignite coal – A form of coal often referred to as brown coal, which has the lowest form of heat production. Derived from peat it is also considered one of the most polluting fuels. Mainly used in coal-fired power stations.

Net-zero – A building or community that generates as much energy as it uses (even though the production may not coincide with use in some cases).

Transition Town – A Transition Town, or Transition Initiative, is a grassroot community project that seeks to build resilience in response to peak oil, climate destruction, and economic instability by creating local groups that work on building local resilience.

Wp (Watt Peak) – An expression of peak electrical output in Watts.

kW – The Kilowatt is equal to one thousand watts – which might power 100 compact fluorescent light bulbs or 10 incandescent ones.

kWp (Kilowatt Peak) – An expression of peak output in Watts, i.e. 1,000 Watts.

kWh – The Kilowatt-hour is a unit of energy equivalent to one Kilowatt of power expended for one hour.

MW – The Megawatt is a unit of energy equivalent to one million Watts.

MWh – The Megawatt-hour is a unit of energy equivalent to one Megawatt of power expended for one hour.

GW – The Gigawatt is a unit of energy equivalent equal to one billion Watts or 1,000 Megawatts.

TW – The Terrawatt is equal to one trillion Watts or 1,000 Gigawatts.

TWh – The Terrawatt-hour is a unit of energy equivalent to one terrawatt of power expended for one hour.

kWth – Kilowatt thermal output.

Resources

USEFUL WEBSITES

100% Renewables
100% RES Communities www.100-res-communities.eu/eng
Community Power Report www.communitypowerreport.com
Global 100% renewables www.go100re.net
REScoops – European federation of grassroots initiatives, support the existing
    and new REScoops across Europe www.rescoop.eu
Solutions Project (U.S.A) www.thesolutionsproject.org

Europe
Community Energy England www.communityenergyengland.org
Community Energy Scotland www.communityenergyscotland.org.uk
Community Energy Wales www.communityenergywales.org.uk

Bath and West Community Energy www.bwce.coop
Centre for Alternative Technology www.cat.org.uk/index.html
Energy4All www.energy4all.co.uk
Low Carbon Hub (Oxford) www.lowcarbonhub.org
Ovesco www.ovesco.co.uk
Plymouth Community Energy www.plymouthenergycommunity.com
Repower London www.repowering.org.uk
Sharenergy www.sharenergy.coop
Westmill Solar www.westmillsolar.coop

DBDH – Denmark's leading district heating organization www. dbdh.dk
Eco-Municipality of Gotland www.gotland.se/eco
EWS Schoenau www.ews-schoenau.de/homepage.html
Feldheim – the energy self-sufficient village www.neue-energien-forum-feldheim.de
German Energy Transition www.energytransition.de
Samso Energy Academy www.energiakademiet.dk/en/om-energiakademiet
Solar District Heating www.solar-district-heating.eu/SDH.aspx

USA
Community Power Network www.communitypowernetwork.com
Institute for Local Self-Reliance – energy http://ilsr.org/initiatives/energy
Rocky Mountain Institute www.rmi.org
Solar industries Association www.seia.org

GRID Alternatives www.gridalternatives.org
Lakota Solar Enterprises www.lakotasolarenterprises.com
Local Power www.localpower.com
Marin Clean Energy www.mcecleanenergy.org
Sonoma Clean Power https://sonomacleanpower.org

Australia
Coalition for Community Energy www.c4ce.net.au
Fund Community Energy www.fundcommunityenergy.org
Enova Energy www.enovaenergy.com.au

Japan
Houtoku Energy www.houtoku-energy.com/project
Institute for Sustainable Energy Policies www.isep.or.jp/en

Global South
Azuri www.azuri-technologies.com
BBox www.bboxx.co.uk
D Light www.dlight.com
Global Alliance for Clean Cookstoves www.cleancookstoves.org
Global Off-Grid Lighting Association (GOGLA) www.global-off-grid-
    lighting-association.org
Grameen Shakti www.gshakti.org
Infrastructure Development Company Limited (IDCOL) www.idcol.org
M-Kopa Solar www.m-kopa.com
Practical Action www.practicalaction.org
Solar Aid www.solar-aid.org
Sunny Money www.sunnymoney.org
Village Infrastructure www.villageinfrastructure.org

BOOKS

Power From the People; Greg Pahl; Chelsea Green Publishing Company, 2012.

Real Goods Solar Living Sourcebook – Your Complete Guide to Living Beyond the Grid with Renewable Energy Technologies & Sustainable Living; John Schaeffer; New Society Publishers, 2014.

Reinventing Fire – Bold Business Solutions for the New Energy Era; Amory B Lovins; Chelsea Green Publishing Company, 2013.

The Home Energy Handbook – A Guide to Saving & Generating Energy in Your Home & Community; Allan Shepherd, Paul Allen & Peter Harper; Centre for Alternative Technology Publications, 2012.

The Solar Century – The Past, Present & World-changing Future of Solar Energy; Jeremy Leggett; Profile Books, 2009.

Sustainable Energy – Without the Hot Air; David J C MacKay; UIT, 2008.

The World We Made; Jonathon Porritt; Phaidon Press, 2013.

Biomass
Planning & Installing Bioenergy Systems – A Guide for Installers, Architects & Engineers; The German Solar Energy Society; Earthscan, 2005.

District Heating
District Heating & Cooling; Svend Frederiksen & Sven Werner; Studentlitteratur AB, 2013.

Geothermal
Geothermal Energy – Utilisation & Technology; edited by Mary H. Dickson & Mario Fanelli; Earthscan, 2003.

Heat Pumps
Heat Pumps for the Home; John Cantor; Crowood Press, 2011.

Hydro
Guide on How to Develop a Small Hydropower Plant; European Small Hydropower Association, 2004; free English download at: www.esha.be/fileadmin/esha_files/documents/publications/GUIDES/GUIDE_SHP/GUIDE_SHP_EN.pdf

Micro-Hydro Design Manual; Adam Harvey; Practical Action Publishing, 1993.

Photovoltaic
Electricity from Sunlight – An Introduction to Photovoltaic; Paul A Lynn; Wiley, 2010.

Pico-solar Electric Systems – The Earthscan Expert Guide to the Technology & Emerging Market; John Keane; Routledge, 2014.

Solar Thermal
Solar Domestic Water Heating; Chris Laughton; Earthscan, 2010.

Tidal
Ocean Energy; Roger H Charlier & Charles W Finkl; Springer-Verlag, 2009.

Wind
Wind Power; Paul Gipe; James & James, 2004.

Index